AN OUNCE OF PREVENTION

RAISING AND FEEDING ANIMALS NATURALLY

Alethea Kenney, B. S., D. Vet. Hom

Publisher: Boreal Balance, LLC
Publisher Address:

Shevlin, MN 56676
Name: Alethea Kenney

ISBN: 978-0-9980-9610-0 (sc)
ISBN: 978-0-9980-9611-7 (e)

Library of Congress Control Number: 2017902666

Because of the dynamic nature of the Internet, any web addresses or links contained in this book may have changed since publication and may no longer be valid. The views expressed in this work are solely those of the author and do not necessarily reflect the views of the publisher, and the publisher hereby disclaims any responsibility for them.

All Images Credit Alethea Kenney

Rev. date: 03/15/2017

ACKNOWLEDGEMENTS

My eternal gratitude to my editor, Cara Sawyer for her patience and careful work, to my animals for teaching me and most importantly, to God from whom all blessings flow.

CONTENTS

PROLOGUE

The more I read about and see how plants and animals are raised today, the more I am convinced that the health epidemic in the United States (and spreading worldwide) has at its root the mega farm. Farming methods have changed substantially in the last 100 years. We have entered the industrial revolution with a fervor that leaves us breathless, literally. Pollution, degradation of soils and water, and even erosion of our own morality have resulted in soils unable to sustain life, soils barren of the trace nutrients and beneficial relationships necessary for healthy plants, healthy animals and healthy people. The bottom line of any decision regarding animal care is now financial rather than being the moral obligation we have to respect the lives in our care. Advertisements and lobbying groups convince us and our government that the respectful methods of our ancestors are no longer valid.

Rather than work to restore what we have lost, companies convince us that more chemicals, further manipulations and pollution will mean we can raise food in spite of the desecration of our world. Once, animals were treated as living creatures worthy of respect for the products and services they provided humankind. Plants were revered for the healing and nutrition they gave. Now, animals are a commodity, their lives and health are treated with disdain. Conventional farming rationale tells us that more animals are better, even if the animals themselves suffer due to overstocking. This includes factory farms, but many smaller farmers, even those who practice certified organic methods, fall into the trap that more is better. Flocks and herds deplete pastures, farmers cannot afford in time or money to treat their animals properly and individual animals become lost in the crowd. Genetic manipulations now include splicing genes of plants and animals into unrelated species to create abominations. The Bible contains a caution against just such a horror: "Keep my decrees. Do not mate different kinds of animals. Do not plant your field with two kinds of seed. Do not wear clothing woven of two kinds of material." (Lev. 19:19 NIV).

Animals are fed their own species back as supplements. Horrific! Bonemeal is now included as a calcium source for herbivores. Dog and cat food protein sources now contain deceased companion animals in addition to the various ill and chemically-filled farm animal protein products. A cow would never choose to eat one of its own and dogs will not eat their own pack mates. We force animals to be cannibalistic and then wonder why their brains become infected.

Our disconnect with the world around us and unnatural connection to technology has created children who no longer know the names of the plants in the flower beds, gardens or sidewalk cracks (or perhaps even realize there are plants there). Adults know them only as weeds to be eradicated with whatever means necessary. Animals and the way of life connected to them is being legislated out of existence. Ecosystems are disrupted and cycles no longer work to keep so-called harmful species in check. We are living Rachel Carson's "Silent Spring" and most of us now know no other way of life with which to compare ours.

The disrespect to the other living creatures with which we share this world, and to the Creator who lent them to us, is appalling. We reap what we sow and we are sowing a nightmare. This must stop if we are to survive as a people and if our earth is to heal.

All is not lost, however! There is still time to change and to turn again to a sustainable way of life. In order to do so, we must recognize there is a problem and understand how to address it. The following pages give information on how animals live, thrive, reproduce and die. By better understanding what makes a healthy animal, and even what that health looks like, we can work to create a farm where animals can be healthy.

Common Sense

Since this book deals specifically with problems and imbalances in animals, it should go without saying that people perusing this book for more than entertainment should absolutely have a veterinarian with whom to consult! In fact, it can be crucial. Even the most experienced farmer or animal lover will run into situations where they either need help, or frankly, have no idea what is wrong. Being able to consult with a veterinarian to have tests done, get advice on diagnosis or, if needed, perform surgery is only prudent. It never hurts to get a second opinion, especially if yours is the first one. You can always choose to treat your animal in a natural way or to do conventional triage care immediately and use natural methods to restore and maintain health afterward. Please use common sense when dealing with animals and be willing to consult with a veterinarian or practitioner when necessary.

CHAPTER 1

THE ROOT OF HEALTH

The digestive system is the basis of health and immunity. Starting before birth, when the population of bacteria and protozoans are introduced into the digestive system, the process of taking in food and breaking it apart into usable nutrients for the body to build, repair and reproduce is the basis for life. Immunity starts at birth for mammals and the immune system grows and learns as the young animal grows and is exposed to potential disease-causing organisms. The link between a healthy digestive system and a healthy immune system is very close and is as important for animals as it is for people. Just as the digestive system and nutrition are the base of immunity and health in people and animals, so the soil is the basis of health and nutrition to the world. How we live and raise our food affects how our world works and how our bodies are able to grow, stay healthy, reproduce and be creative. There are already some excellent books available about restoring soil fertility organically and biodynamically, raising crops for food and sustainability and even raising animals naturally.

This book is an attempt to help people make different choices about care of their animals from the ground up. By understanding the importance of digestion and how food affects health for life, people can make good decisions about proper, species-specific feeding for many animal species. Because chemicals also affect health, this book contains information on choosing alternative, natural treatments where necessary to restore health. These natural treatments do not negatively impact the animal (remember, first do no harm) or the environment in which it lives. Chemical treatments not only have side effects in the animal, they have the potential to damage the delicate ecosystem by introducing antibiotics and anthelmintics into soils where animals defecate.

Animals and humans have an intricate relationship on this planet. Wild animals are valued for their beauty, sport opportunities (bird watching, hunting, safaris, etc.) and food sources. Throughout the history of the world, indigenous cultures relied heavily on wild game for their food, clothing and shelter. Humans even used animals as part of their religious beliefs and ceremonies, art and culture. Domestic animals, bred to work closely with humans, began to replace wild game for the same functions. In modern society, domestic animals can be divided into roughly two categories: those that provide companionship and work closely with humans as guards and pets, and those raised for food or products, such as sheep, goats, cattle and fowl. Among these types of animals, there are three categories: herbivore, omnivore and carnivore.

The diets and nutrition of each type of animal differs. In order for them to survive, thrive, grow and reproduce, their diet must meet nutritional needs. Proper diet is critical to health, immunity and reproduction in all animals. Domestic animals, unlike their wild counterparts, cannot roam and forage for whatever food they need to meet their nutritional needs. Domestic animals are usually partially or completely dependent on humans to provide what they need for food and fresh clean water.

Understanding how animals take in food and digest it to provide nutrients for growth, maintenance and reproduction allows an understanding of what foods are proper for each species and the importance of providing foods that also maximize digestion.

Improper and incomplete digestion leads to build-up of metabolic toxins in the body, inflammation and becomes the root of susceptibility to all disease. The increasingly toxic environment we live in also increases toxic burden on the body, making proper digestion and elimination of toxins even more important for health. First, the digestive system is covered. Then elimination of toxins and eliminative channels will be discussed, with information on supporting proper organ function, helping the body remove toxins and rebuilding healthy tissue.

Changing Ideas but Not Changing Needs

Animals have not changed appreciably over the last thousand years and their dietary and lifestyle needs have not changed either. How we keep and feed our animals, however, has changed substantially in the last fifty or so years. Animal feeding methods have been researched and written about for many many years. Soil science has grown to include the toxic chemicals, manipulation of plant genetics and other unsustainable methods used by corporations, rather than family farmers who know the land and have a vested interest in conserving it. Pet food has become a multibillion-dollar industry that, like pharmaceutical companies, courts veterinarians and provides product marketing to encourage veterinarians to recommend certain brands. Marketing does not always equate to the best product for the health of the animals!

Books and research done before the advent of the chemical revolution focus on nutrition from the standpoint of more natural sources and methods of keeping and feeding animals.

Vaccination

This will be addressed in a later volume but I will make a few comments here. If healthy animals do not maintain high levels of parasites or disease, how necessary are vaccinations and do they provide the protection that is claimed by pharmaceutical companies? This is a delicate discussion because so much of the information available is propaganda rather than actual, verifiable data about the benefits, and especially the side effects, from vaccination. In general, if the animal in question manifests a disease supposedly prevented by vaccination, the first question should be, "What is the underlying imbalance that allowed such a disease to take hold and how can I work to correct that?" I choose not to vaccinate my animals except for those vaccines required by law. I did this because of the information and research I have found. The volume on diseases will provide references so that interested persons may find information and can make wise, informed choices.

Effects of Stress

If an animal is experiencing stress, probiotics is a good adjunct therapy but it is not the whole story. Stress can reduce immune function and increase susceptibility to diseases, and beneficial populations of bacteria and protozoans can be depleted during times of stress.

Minimization of stress is paramount to health in all animals. Livestock should be raised in as natural a setting as possible and emotional health of domestic pets should be considered. Dogs are naturally social animals and may need a companion or more time with their owners. Cats are more solitary and may be territorial with multiple cats in one house. Flower essences, homeopathy, management and proper diet all work to help reduce stress and keep animals emotionally healthy.

Use of probiotics can re-establish some populations of bacteria but it is important to remember that no probiotic can completely restore a depleted digestive system. There are many strains of both bacteria and protozoans that are part of an intricate and necessary symbiotic relationship in the digestive tract of healthy animals. Maintaining and restoring the environment that allows these beneficial organisms to thrive is as important as reintroducing a few strains after stress, illness or accident because no probiotic supplement can contain all the species necessary for good health.

Humane Care

Healthy animals do not maintain high levels of diseases, parasites or other problems. Health depends on food and lifestyle, including mental and emotional health. Animals must be treated humanely and kept in a manner consistent with their needs as a species, both physical and emotional. Animals under stress will have digestive problems and subsequent nutrition-related imbalances. They will become unhealthy, unthrifty, unhappy and unmanageable. It is always worth the time and effort to design pastures and living arrangements that will allow the animals to live as they were intended.

Living on the Land vs. Living with the Land

The land is an extension of the animals that live there and domestic animals are not different than wild animals in their need for a healthy environment. The farm should take into consideration the parts of a thriving ecosystem and how each animal and plant has a role to play in keeping the land healthy.

Something that is overlooked more and more in modern society is the human need to live in a healthy environment. Western culture (and now many others) has at its root the need to dominate the environment rather than live in communion with the land. This attitude changes how humans interact with the land and animals, leaving them more as visitors or worse, destroyers, rather than as one more piece of a multi-faceted world. Some people believe humans should never be a part of some lands and others believe humans must subdue lands. The reality is that humans have always been a part of the world and the land, I would argue, an integral part of an intricate system. Although now we no longer completely depend on the land at our doorstep for all our food and shelter needs, we are still completely dependent on the earth for our life. Those people who choose to continue to keep animals as companions, farm animals or working partners now must choose to live in harmony with, rather than in antagonism with, the animals and land. Everything from organic and biodynamic methods in restoring soils to gentle training methods for animals that depend on a mutual, respectful and positive relationship are part of the movement to include humans in their role as an integral part of this world we live in. Caretakers of the earth strive to restore what has been lost and nurture the relationship between humans and earth, plants and animals. At its root, this restoration must include a nurturing and love for fellow humans as creatures important to the well-being of the earth and the Creator. From this beginning, education becomes the basis for helping people understand how important their contribution is to the healthy functioning of the world and the animals in it.

Another concept related to this topic involves choosing breeds and species of animals for farms that complement the farm and its management goals and fit well into the physical environment. If the farm is seen as a smaller ecosystem sphere, the animals on the farm should be a part of this relationship with the earth and the people living with it. A thriving and living farm needs to contain animal species and breeds that work well together and in the climate, weather and pasture conditions of that farm. Diversity can be the key to health for all components of this system. No animal or plant lives in isolation but becomes an integral part of a much larger system.

As an example, I live now on a farm in northern Minnesota at the confluence of the hardwood, coniferous and prairie ecosystems. Soils can be "poor," forest soils are notoriously poor in nutrients in top layers. Trees reach into subsoil for their nutrients. A thriving forest gives the impression the soils must be rich to support such a growth but, in fact, if the forests are cleared, topsoils are not rich at all. The climate can be extreme, winters are long and cold, almost unbelievably so with lows reaching -50 °F (-45.6 C) ambient air temperature on occasion, and summers generally short, cooler and wet. Some people here choose to keep cattle. It is important to realize if you do, that cattle must be fed differently in cold weather since they do not have woolly coats for protection. Sheep, on the other hand, do have protection from the elements and do very well in cold harsh winters. But not all sheep breeds are the same. I've carefully researched breeds to find and breed for animals that need little to no intervention or supplemental feeds. In addition to animals, birds can be a good complement. In this area, wet pastures promote snails that carry liver fluke and meningeal worm (More about those in Volume 2). Ducks loves snails! And ducks are also fairly winter-hardy, produce eggs, meat and feathers while removing unwanted insects and mollusks from pastures. Pigeons are a forgotten livestock animal that can provide meat, eggs and feathers and have the advantage of tolerating cold while continuing to raise young. While chickens suffer in cold weather and tall combs tend to freeze off, other species can thrive in such a situation. Pay attention to breed differences and choose animals that complement each other.

People living in warm, humid climates will choose different breeds and species to complement their farm goals or modify their goals to be consistent with living with the land and ecosystem in which they find themselves.

A note about breeds: sheep and goats have always been the animals people kept when they could not afford to feed cattle and had only marginal lands for grazing. Sheep and goats turn coarse grazing and browse into meat, fiber and milk and have the potential to do this without supplemental feedings or interventions. Unfortunately, we have now bred most of the primitive heritage breeds of these animals to require expensive inputs of grain, time and emotion in order to produce less than their ancestors did. Do not be fooled by the rosy picture of breeds of livestock that breed out of season and produce so many young they cannot raise them without extreme intervention. In cold climates, lambs born very early in spring must be able to get up, nurse and stay by their mothers. Mothers must have good instincts. Without this, lambing has to be done in a building with heat. This is not necessary since, historically, sheep and goats have been able to give birth and raise up young in extreme environments without intervention. These situations require not only proper minerals (low selenium yields weak lambs unable to stand or nurse at birth and at high risk for hypothermia and death shortly after birth) but also a conservation of genetics that give animals an ability to thrive under extreme environmental conditions.

The more you do to an animal, the more you will have to do for that animal. This phrase applies to all livestock species but not companion animals. Even in companion animals, though, these concepts should be considered when breeding. Do not breed those animals that cannot reproduce, thrive and survive without high input! The more things we buy and use on and in our animals, the more things we will have to continue to do to keep them healthy. We are selecting for animals that survive only with more and more inputs, instead of selecting for the animals that did not need our attention, money or supplemental feeds.

Many breeders of heritage breeds now feed grains and give all kinds of medications, dewormings and supplements. This is inadvertently selecting for genetics that require all those inputs and is an expensive way to ruin a breed that has served our ancestors for centuries. I have not talked to anyone who does this on purpose. No one goes out to the pasture and thinks, "If only I could create a breed that needed to be dewormed every other week all summer and could only lamb with my assistance after feeding pregnant animals expensive feeds." People do not understand what true health is and how to achieve that in a population of animals. Hopefully all of us can start to give this more thought as we set up our farms and breeding programs. More educational programs should focus on the full scope of sustainability as it relates to breeds and management.

Keep in mind, there is nothing whatsoever wrong with pet animals, be they sheep, dogs, ferrets or horses. Pets, fiber animals and companion animals are in a different category completely. These animals serve us as friends, family, emotional support, fiber producers and give us the pure enjoyment of interacting with and seeing beautiful animals. Their health is determined by the health of their ancestors and that may have been determined by someone's misunderstanding of good breeding practice or the vagaries of a feral mate but they are worth coddling because they are here. And we enjoy showering them with attention and care. We spare no expense in giving them the best. There's really nothing wrong with this at all. Realize that these animals may come with health problems that take a lifetime to unravel (or are not completely correctable) and plan to treasure them for who they are.

The following information will help those who have companion animals and pets, those who have breeding operations and those who want to learn more about how good nutrition and digestion promote health and growth in all animals (and people). And I sincerely hope this book will guide people to making healthier choices for their animals, themselves and the planet.

In this book, organ systems, minerals, vitamins and health are broken apart into pieces but this is only a small (and not particularly useful) part of the equation. The body and the land must be seen as whole and interconnected. In order to understand the whole system, however, it can be helpful to understand each part. Just remember that although the discussion concerns pieces and parts, these must be put back together to create a whole (and holistic) view.

Mammals and Their Differing Digestive Systems

Mammalian digestion is diverse, divided mainly into two categories, extreme examples of which are carnivore and herbivore, with omnivore in the middle. This diversity continues into organs in the digestive system. The main organs of the digestive system in all mammals are the mouth, teeth, tongue, pharynx, esophagus, stomach, small intestine and large intestine with accessory organs like salivary glands, liver and pancreas. Most all mammals have the same organs but those organs differ considerably in size.

Teeth include incisors, canines, premolars and molars on each side in most species. Tooth placement is indicated by numbers in a formula given in textbooks that correspond to the number of each type of tooth on one side of the mouth, top and bottom. An example is 0033/4033 for a cow. In this example, top teeth one side are given as the first (or top) portion of the fraction and bottom of one side is given on the bottom of the fraction.

0 incisors, 0 canines, 3 premolars and 3 molars

4 incisors, 0 canines, 3 premolars and 3 molars

As a comparison, the dental formula for a dog is:

3142/3143

Top one side: 3 incisors, 1 canine, 4 premolars and 2 molars
Bottom one side: 3 incisors, 1 canine, 4 premolars and 3 molars

From these differences in dental formulas, it can be seen that dogs have teeth designed to tear into flesh while cattle have teeth designed to pull up plants. Shape of premolars and molars differs between species as well. Carnivore molars are designed for crushing meat and bones, herbivore molars contain sharper ridges that break down plant fibers. Changes in jaw structure between the two species illustrates how carnivore jaws move up and down, herbivore jaws (and more particularly, ruminant jaws) can move laterally, allowing cows and other herbivores to better grind up plants.

The differences between species become more striking further down the digestive system. For instance, most of the digestion in the horse takes place in the cecum and large intestine while in the dog, this takes place in the small intestine.

Digestion in the small intestine consists of enzymes that can break down carbohydrates, fats and protein. Cellulose (plant material) digestion does not occur in the small intestine and since carnivores, such as dogs, do not take in much plant material, they depend almost completely on the small intestine for nutrition.

Importance of Understanding Digestion for Proper Feeding

Why is it important to note differences in digestion from one species of animal to another? Knowing how each species of animal digests its food allows for careful decisions on the best and most natural foods and mineral sources needed to meet the needs of and support proper digestion in each species. Improper feeding leads to disease, malnutrition, emotional problems and reproductive issues in all species subjected to the poor feeding plan.

Example Digestive System Differences by Species

Seeing the differences in the digestive system of animal species helps explain why some species can digest plant material and some species really cannot. The chart below compares the length of the small intestine in different species, including herbivore, carnivore and omnivore.

Length of small intestine

Herbivore:

Horse 24.5 yds (22.4 m)
Ox 50.3 yds (46 m)
Sheep/goat 28.4 yds (26 m)

Omnivore:

Pig 6 yds (5.5 m)

Carnivore:

Dog 4.5 yds (4 m)

Ratio of body length to intestine:

Herbivore:

Horse 1:12
Ox 1:20
Sheep/goat 1:27
Rabbit 1:10

Omnivore:

Pig 1:14
Dog: 1:6

Carnivore:

Cat 1:4
Ferret 1:5

(Swenson, 1984, p. 263, 271; Fox, 1988, p. 43; Curtis, 2011, p. 4; Reece, 2009, p. 372)

A closer look at how digestion takes place in a representative species of herbivore, carnivore and omnivore will give more insight into how management and feeding will need to change from one species to another and why each animal species is adapted to thrive on a particular diet.

CHAPTER 2

RUMINANT HERBIVORE DIGESTION

Ruminants

Domestic Examples: Cattle, sheep, goats

The stomach in ruminants is usually divided into four compartments. Camelids, like llamas, alpacas, guanacos and vicuñas, have three compartments and a few different features.

Non-Ruminants

Domestic Examples: Horses

The stomach is simple, not divided.

These animals are referred to as having monogastric, or simple stomach digestion since their stomachs have only one compartment, although this may be utilized differently from top to bottom.

Ruminant Types

Ruminants are divided into three types based on feeding habits and types of plants eaten. The plants referred to as browse include shrubs and trees, plants that are grazed include grasses and forbs (flowering plants). Browsers (called concentrate selectors in nutrition books), such as white-tailed deer, choose plants that are very digestible, high in starches, proteins and fats but low in fiber. They do not ferment cellulose well and cannot break down cell walls of plant material as well as other types of ruminants so generally choose browse rather than grass-type plants. Grazers, or grass/roughage eaters, are able to digest cellulose fibers in plants well and include animals like sheep.

Intermediate type feeders will eat forage and more concentrated plant material, depending on season and availability and include goats and reindeer. They will choose browse when it is available but will eat and are able to digest a grass-based diet (NRC, 2007, p. 5).

Because it takes time for bacterial colonies in the digestive system to adapt to changes in feed, any changes in diet should be introduced slowly. This fact is one reason why attempts at artificial feeding of elk overwintering in some states may result in elk starving anyway. The bacterial colonies in elk digestive systems cannot change quickly enough from digesting browse that is available in winter to digesting richer grass and alfalfa hay. The same holds true of domestic herbivores. It is important to segue into seasonal diets gradually so the digestive system bacteria and protozoans have time to adapt to the new foods. Mimicking nature provides the best way to support ruminant digestive systems.

Organs of the Small Ruminant Digestive System (with Goat as an Example)

The interior hollow tube that is the digestive tract is technically outside the body, a closed system that is separated from the rest of the body by a lining of epithelial cells. This allows toxic substances to be separated from nutritious ones without causing damage to the rest of the body. The epithelial cells absorb nutrients and secrete enzymes needed to digest sugars and peptides (amino acid chains) (Swenson, 1984, p. 263). In order for nutrients to be available for absorption, the foods taken in must be digested or broken down. This occurs in the intestine, aided by secretions of the salivary glands, pancreas and liver. Secretions include water and electrolytes, digestive enzymes and bile salts while mucus cells lining the digestive tract excrete acidic or basic substances to maintain the proper environment (Swenson, 1984, p. 263).

The neuroendocrine system controls secretion of digestive substances and the movement of foodstuff through the digestive tract. Absorption of needed nutrients and reabsorption of digestive secretions like water and electrolytes must take place (Swenson, 1984, p. 263). This link between the digestive system and the neuroendocrine system is the reason why digestion stops when an animal is under stress. The body recognizes that protecting life is more important during crises than digesting food. The neuroendocrine system includes glands like the pituitary, hypothalamus, adrenal and thyroid.

Mouth

This is the start of the digestive tract and even here, there are obvious differences between different ruminant species. Lips can differ from species to species. Grazers have broad mouths and short lips, making it easier to take in large amounts of grasses. Browsers have a more narrow mouth that can accommodate particular plant parts and fruits. The goat upper lip is not divided like a sheep's lip, favoring both browsing behavior and grazing behavior (Smith, 2009, p. 377; NRC, 2007, p. 6). In browsing species, the mouth is adapted to pull leaves from thorny branches and strip quality plant material from the twigs of shrubs and trees (NRC, 2007, p. 6).

Tongue

The tongue is a muscle used to move food in the mouth and contains taste buds. Food taken in an unprocessed state allows an animal to distinguish between healthy and harmful foods, but animals have trouble making that distinction with processed foods (Reece, 2009, p. 364). Goats have a tongue that is not prehensile like a cow's. The goat tongue is short, smooth and not easily moved out of the mouth, and the taste buds tolerate bitter tastes much more than sheep or cattle. In grazing species, the structure of palate and tongue allow rejection or continued chewing of plant parts based on taste (NRC, 2007, p. 6).

This ability to detect nutrition and needed substances by taste means herbivores can choose to eat what they need if it is available to them. As noted above, however, feeding concentrates, pelleted feeds or manipulated feeds negates this ability and animals are unable to determine whether the food is healthy and needed or not. Flavor enhancers in concentrated products further manipulate the ability of animals to know what they need. For this reason, I do not assume that most flocks or herds still possess that innate sense, unless they have been raised without additives or concentrates.

Salivary Glands

Saliva, which is produced in the salivary glands, starts the digestive process and helps moisten food. For example, in goats there are four main pairs of salivary glands-parotids, mandibulars, sublinguals and buccals.

Teeth

All ruminants and camelids (llamas, alpacas, vicuñas and guanacos) are missing the incisors on the upper jaw and most species also lack canine teeth on the top. Maxillary (upper) molars and premolars are wider than those of the mandibular (lower). Lateral chewing motion (side to side) helps break up the plant cells (NRC, 2007, p. 7).

Sample dental formula for goat: 0033/4033 (Smith, 2009, p. 378).
Sample dental for horse as comparison: 3133/3133 (Reece, 2009, p. 362).

Chewing motion differs from species to species. Ruminants have a lateral grinding motion for chewing.

The age of the animal can often be determined by knowing when each set of teeth erupts and how they wear over time. Like humans, herbivores start life with milk (or baby) teeth that fall out as the animal matures, to be replaced by permanent adult teeth. As an animal ages, teeth wear and there can be digestive difficulties because of this. Minerals and nutrition play a role in tooth health but time and fibrous diet will eventually wear down even the healthiest teeth. When this happens, if the animal is to be kept healthy, feeding decisions need to be made about offering higher nutrition in a more palatable form.

Esophagus

The esophagus is a tubular organ that carries food from the mouth to the stomach (in ruminants to the rumen). In most goat species, the esophagus is approximately three feet in length. A ventricular (formerly called esophageal) groove in nursing kids allows milk to by-pass the rumenoreticulum and go into the abomasum. Adults can occasionally by-pass the rumenoreticulum during water intake after dehydration (Smith, 2009, p. 378). The rumenoreticulum is the name for the rumen and reticulum when both compartments are considered together. The groove has two sections, reticular and omasal. When the lips of this groove close, milk is diverted directly to the abomasum (NRC, 2007, p. 6).

Stomach

In ruminants, the stomach is divided into four compartments (camelids have three). These compartments are the rumen, reticulum (left half of abdomen), omasum (much smaller than the reticulum and much smaller in goat and sheep than in the cow) and abomasum, the "true" stomach. The abomasum is proportionately larger and longer in sheep and goats than in cows (Smith, 2009, p. 379).

The rumen is where ingested fibers go first after being soaked in saliva and chewed. After digestion by bacteria, protozoa and fermentation, the fibers are regurgitated and chewed again, then swallowed again into the rumen. When particles of fiber are small enough, they pass to the reticulum. Foreign objects that are not digestible settle to the bottom of the reticulum and stay. Fermented particles pass to the omasum where water and nutrients (volatile fatty acids) are removed. The particles are then forced into the abomasum for further digestion with hydrochloric acid (HCl). This compartment is the same as the simple stomachs of carnivores and omnivores.

Goats and lambs are born with an undeveloped rumen but larger abomasum, which changes as the young animal is exposed to a fibrous diet. Full function and capacity of the rumen is reached by age twelve weeks in young kids (Smith, 2009 p. 379).

The reticulum can contract to mix and move food into the rumen, facilitate regurgitation from the rumen to the mouth and expel food and gases into the omasum. The rumen is large and contains a huge quantity of bacteria and protozoa for fermentation. Food eaten is stored here for later regurgitation. Like in the reticulum, there is constant movement of foodstuff. In grazers, the movements cause a stratification or layering of food, which can make sheep more susceptible to bloat than camelids who do not have such layers of food in their rumen (NRC, 2007, p. 7). Papillae line the rumen for absorption of fermentation acids. The rumen and reticulum are close to each other and pass food back and forth often enough that these two compartments are often considered together as one, the rumenoreticulum (NRC, 2007, p. 8).

Grazers have larger rumen volume than browsers and size of rumen/reticulum increases in volume with body weight (NRC, 2007, p. 8). Browsers have a larger omasum and, in intermediate species, rumen/reticulum size varies by season and type of forage. The compartments are smallest in winter (NRC, 2007, p. 8).

Size is also associated with amount of food available. Grazers have a smaller opening into the omasum and process fiber longer in the rumen/reticulum than do browsers. Browsers take in larger amounts of undigestible woody material with the leaves and fruits, rapidly digesting the higher quality leaves (NRC, 2007, p. 9).

The omasum has layers of absorbing lining; little digestion occurs here, but minerals and water are removed (NRC, 2007, p. 9). Food then passes to the abomasum.

The abomasum acts like the stomach of non-ruminant animals. Secretions from parietal glands mean the contents are higher in water as they pass into the small intestine. Browsers and intermediate species have larger, thicker-walled abomasums than grazers, with more HCl secretions (NRC, 2007, p. 9).

Enzyme-Secreting and Accessory Organs

Salivary Glands

The salivary glands secrete enzymes important for digestion, like amylase. Amounts of enzymes vary depending on age and species. As an example, young pigs (omnivores) do not secrete high levels of amylase and levels drop off further as pigs mature. Since amylase is important in starch digestion, it is obvious that pigs do not digest starch well. Ruminants and carnivores do not produce amylase in the salivary glands.

Low-level secretions of gastric enzymes in the newborn mean colostrum is not digested before it can reach the intestine for absorption. After about two weeks of age in pigs, HCl acid secretion increases (Swenson, 1984, p. 265).

Pancreas

The pancreas secretes digestive enzymes and hormones, depending on species and age. Amylase, lipase, trypsin and chymotrypsin are digestive enzymes produced by the pancreas. Young animals start to increase production of these enzymes at about weaning age (Swenson, 1984, p. 265). Hormones like insulin and glucagon are also produced in the pancreas. Insulin allows cells to be able to use glucose. Glucagon causes the liver to release glucose, that has been converted from stored glycogen, into the blood stream. Insulin acts to help glucose cross cell membranes but excess glucose can cause an increase in insulin production, leading to cells that become desensitized to the insulin and no longer respond by absorbing the blood glucose. This process is termed insulin resistance and is the first step in Type 2 diabetes in humans and metabolic disorder that leads to obesity, truncal fat, slower metabolism and other disorders. It is linked also to the HPA axis (the neuroendocrine system discussed later that includes the hypothalamus, pituitary and adrenal glands). In horses, if cells become resistant to insulin, because of overproduction in response to a diet high in starch, the pancreas produces more and more insulin. High blood sugar is rare in horses, unlike in humans where Type 2 diabetes results from insulin resistance. Instead, horses gain weight and become susceptible to issues like laminitis. Laminitis relates closely to low magnesium, though, also an important factor in support of the HPA axis feedback system and metabolism (Getty, 2013, p. 2).

Small Intestine

The small intestine is seventy-seven percent of the length of the digestive system, made up of three sections: duodenum, jejunum and ileum. The bile duct opens into the duodenum and pancreatic secretions also enter duodenum. Between species there is much variation in length of the small and large intestines. Villi lining the walls of the small intestine are where most nutrients are absorbed and digestion takes place (NRC, 2007, p. 10).

The pancreas, gall bladder and intestine itself all secrete into the small intestine to break down fats, proteins and starches into fatty acids, amino acids and sugars, respectively. The remaining undigested portion passes into the large intestine (NRC, 2007, p 10).

Grazers have a small intestine that is long relative to body length. Browsers have a shorter small intestine relative to body length. Both are still much longer compared to carnivores and omnivores (see the chart above showing ratios) (NRC, 2007, p. 10).

Liver

The position of the liver in the body differs by species but it is always right behind the diaphragm (Reece, 2009, p. 378). In addition to its role in producing and transporting lipoproteins to the body and storing excess glucose, the liver breaks down proteins and stores fat-soluble vitamins (A, D and K) and B vitamins. The liver is also the great detoxifier of the body, filtering the blood through the hepatic portal system. Blood comes into the liver from the portal vein (from stomach, pancreas, spleen and intestines) and the hepatic artery to be passed through the sinusoids where it is detoxified and returned to the venous system. The liver also filters out foreign matter and tissues from the blood (Reece, 2009, p. 379). In conjunction with digestion, the liver produces bile salts that are stored in the gallbladder in most animals to be released into the small intestine. The horse has no gallbladder, but bile continuously drips into the small intestine in response to fat digestion (Reece, 2009, p. 378; Getty, 2013, p. 24). The liver also stores minerals not needed immediately by the body, and consequently, liver tissue analysis can provide a general picture of how an animal is utilizing minerals from the diet. Most tissue sampling is done from necropsy since liver biopsy is invasive and expensive.

Large Intestine

No important nutrient absorption takes place in the large intestine but absorption of water and inorganic ions, like minerals, takes place here (NRC, 2007, p. 10).

Cecum: The cecum makes up two percent of the length of the digestive system (Smith, 2009, p. 379). Contents of the small intestine enter the large intestine at the cecum (ileocecal junction) in the horse, at the colon (ileocolic junction) in the dog and cat or ileocecocolic junction in ruminants and pigs (Reece, 2009, p. 374).

Colon (ascending, transverse, descending) and Rectum: The elongated ascending colon and the shorter descending colon lead to the rectum. The colon and rectum are twenty-one percent of length of the digestive tract (Smith, 2009, p. 379).

The cecum and colon of non–ruminant herbivores continue fermentation. In ruminants, the forestomachs (rumen/reticulum and omasum) are where fermentation takes place and when enzyme digestion occurs and the bacteria and protozoans of digestion do their work. In non–ruminants, enzyme digestion comes before fermentation and the bacteria and protozoans are not available at that stage. The cecum continues fermentation (unlike in carnivores where it is poorly developed or lacking entirely, as in the case of the ferret).

The cecum empties into the colon, which has three parts. All animals have transverse and descending colons but the dog and cat have an ascending colon between cecum and transverse colon, while horses, ruminants and pigs have a different part that corresponds to the ascending colon. In pigs and ruminants, it is the coiled colon (ansa spiralis), in horses it is the large colon (ventral and dorsal) (Reece, 2009, pp. 374-375).

Omentum and Mesentery: The omentum is a membrane that encloses the upper part of the digestive organs, securing the stomach to the abdominal wall. The mesentery covers and secures the lower part of the digestive system, including intestines. Goats lay down fat here in much higher amounts than cattle or sheep, an adaptation for environments with less feed. This may present a problem if goats are overfed in early pregnancy and lay down too much fat. The fat will take up space that is needed for food capacity and can contribute to pregnancy toxemia (Smith, 2009, p. 379).

In goats, food can be in the digestive tract over twenty-four hours (Tisserand, et al. p. 48).

Cud-Chewing

In ruminants, contractions in the esophagus bring up a bolus of food for re-chewing. This allows the food particles to be more thoroughly broken down, and more food can be packed into the rumen. This increases the surface area of the particles for better digestive activity (NRC, 2007, p. 10).

Fermentation in Ruminants

Many species of bacteria and protozoans exist in symbiosis with the animal and help ferment foods like carbohydrates (cellulose and hemicellulose in a ruminant) into energy, synthesize the B vitamins and vitamin K and convert non-protein nitrogen into protein (NRC, 2007, p. 10). Fermentation takes time and fiber: herbivores have digestive systems developed to maintain larger fibers and large surface areas where microbes can do their work. Bacteria and protozoans are not limited to only the stomach, but are found in the esophagus and occur in various colonies and quantities all along the digestive tract. The environment is always changing slightly but is also fairly stable to support bacterial and protozoal health (Swenson, 1984, p. 340).

In ruminants and camelids, fermentation occurs before the acidic portion of the digestive tract, but in other species, this process occurs after exposure of foodstuff to the acidic digestion (Swenson, 1984, p. 340). This allows ruminants and camelids to absorb beneficial volatile fatty acids both in the rumen and in the small intestine and ammonia can be converted by bacteria to protein for absorption in the small intestine (Swenson, 1984, p. 340). The ammonia conversion allows manufacturers to make urea blocks as protein sources for ruminants.

Rumen pH is usually 5.5-7.2 and 100.4 °F to 107.6 °F (38-42 C), anaerobic (without oxygen), and continuously changing with addition of secretions (like saliva) (NRC, 2007, p. 10).

Camelids

Camelids (llamas, alpacas, vicuñas, guanacos) have three parts to their stomachs rather than four, like the other ruminants. Referred to as compartments C1, C2 and C3, C1 has the largest volume and is where fermentation takes place (like the rumen/reticulum of the ruminants). Camelid C1 has saccules lined with glandular epithelium. Other differences here include greater absorption of volatile oils and a higher pH in the fermentation process. The back part of the compartment is more like the abomasum of ruminants. Camelids also have greater gut motility and the mixture of foodstuff is more homogenous and not layered as it is in a ruminant. Otherwise, digestion in the stomach is similar although llamas pass food more slowly through the system, allowing better digestion of poorer quality foods (NRC, 2007, p. 9).

Natural Diet Considerations

As can be seen above, the ruminant digestive system is designed to work with large amounts of fibrous plant material. Without this plant material, the rumen cannot function properly to bring nutrients, vitamins and minerals into the body for growth, health and reproduction. Ruminants can tolerate toxins in the diet better in some cases than other animals can because the digestive system in ruminants breaks down and destroys the toxins before they reach the small intestine for absorption. Diets unsuited to a herbivore, however, may cause an upset in the bacteria and protozoans in the digestive system, leading to other problems and diseases.

In many cases, ruminant bacteria populations can adapt to gradual changes in diet but sudden changes may cause toxicity and death.

Proper digestive function is also necessary for production and absorption of vitamins, absorption of minerals and other nutrients. Without these available in the diet, ruminants cannot survive. Soils depleted in minerals cannot provide minerals for uptake by pasture plants and therefore these minerals are not available in the diet and must be provided separately. Many farms work toward improving soil health and adding back in lost minerals. Farmers renting land or with small holdings who cannot afford to do a lot of improvement, however, must rely on supplemental minerals for their animals. See Minerals below for more information on this topic.

Access to pasture or wooded areas with a diversity of plant species allows ruminants to choose the type of forage they need. Different species of plants contain different chemicals and may provide medicinal action. Shrubby species and young tree leaves may contain as much protein as alfalfa during certain stages of growth and also contain essential oils, tannins and other constituents that benefit health. Much research is done on the importance of diet for preventing stress from parasite infections, and access to a wide range of plant species ensures ruminants can choose what they need. Tannins (although potentially toxic in high amounts, especially if fed before rumen bacteria can adjust) help kill parasites. Browsing higher off the ground breaks up parasite life cycles by keeping animals from ingesting larvae that climb up grass blades.

Rabbit Digestion

Rabbits are herbivores and similar to sheep and goats with some notable differences in their digestion. They have a body to small intestine ratio of 1:10, similar to other herbivores and closest to the horse.

Like other mammals, digestion starts in the mouth with salivary gland secretions and mastication (breaking up of food by chewing). Food then travels to the stomach, where enzymes and HCl continue to break food into its nutrient components. Food then moves to the small intestine where enzymes from the pancreas continue the digestive process. The small intestine is where absorption of nutrients takes place and connects to the cecum, where more enzymes and bacteria continue the job of digestion. Cellulose and other plant material that is otherwise not able to be broken down by mammals is processed in the cecum. Undigestible material leaves the cecum to enter the colon for removal from the body; however, in rabbits, materials that are still digestible but need further breakdown are passed as night feces or caecotrophs. These night feces are re-ingested by the rabbit as they are passed from the anus (coprophagy). This re-ingestion of partially digested material serves the same purpose as cud chewing in other herbivores and allows the rabbit to make very efficient use of food that would otherwise be undigestible (Kilfoyle & Samson, 1996, pp. 101-102).

See Chapter 4: Rabbits for more information.

Problems Related to Dietary Change or Mismanagement

Lactic Acidosis

Overfeeding of grain means bacteria and protozoans shift from a mostly gram-negative population to a mostly gram-positive population that produces lactic acid during fermentation of carbohydrates. The pH may drop and the normal and necessary bacteria cannot survive, leaving only lactobacilli. Ethanol accumulates, leading to bloat (Swenson, 1984, p. 347).

Overgraining can also produce laminitis, diarrhea, digestive upset and pain, dehydration, weakness and death (Swenson, 1984, p. 352). Excess grain increases the need for vitamin B12 and consequently also increases the cobalt needed to make the B12. These conditions should serve as cautions for the feeding of grains to ruminants.

Nitrate/Nitrite Poisoning

Ruminants can normally reduce nitrates to nitrites to ammonia in the digestive system without toxicity. Higher levels of nitrates suddenly added to the diet do not allow for bacterial populations to adjust. Nitrates can come from several plant species that can otherwise be a good food sources for ruminants (Swenson, 1984, p. 347).

Oxalic Acid Toxicity

Oxalates are common in many plants but can occur in high amounts in some plant species. Again, animals that have adapted their bacterial populations to digesting higher oxalate levels (oxalate-degrading bacteria) are able to digest even higher levels.

The above information illustrates the importance of a proper diet and careful management in maintaining ruminant health. Access to pasture, limited or no concentrated carbohydrates (like grain), no pelleted feeds and no genetically modified (GMO) feeds are all important for proper digestion and health. Changes in grazing and diet should be done slowly to allow the digestive system to adapt. Once this adaptation occurs, ruminants can be very efficient at digesting plant materials otherwise indigestible or potentially toxic to other animals.

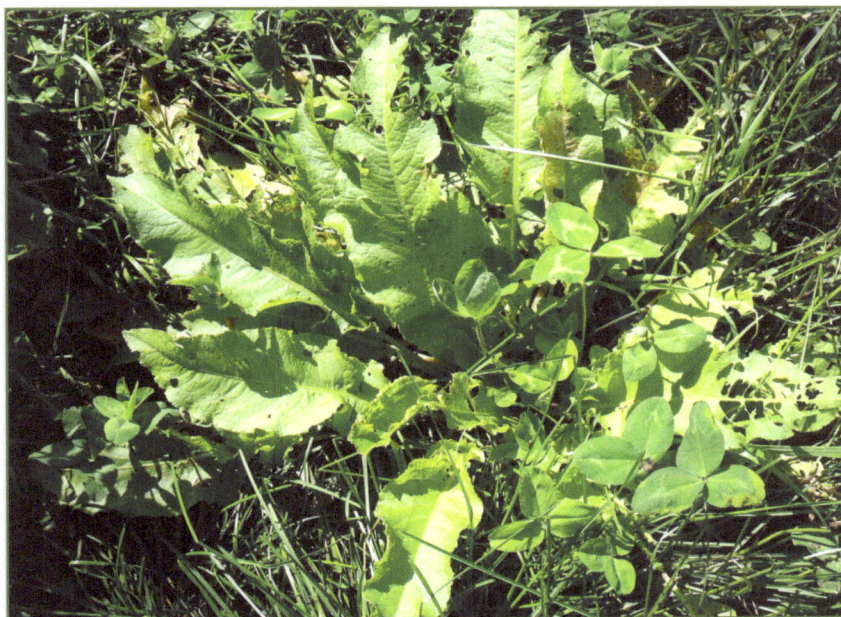

Molasses/Systemic Yeast Infection

A note here about molasses and sweet feeds: molasses, while a potentially healthy addition to food in small amounts occasionally, is a source of sweet carbohydrates that herbivores do not need. The addition of molasses to an already high-grain diet only increases the susceptibility of the animal to parasites and fungal infection. LaMancha goats are notorious for having a tannish yellow discharge from their ears, to the point that some herbalists sell products to treat the ears of these goats on a regular basis. This discharge is yeast! Treating this topically is touching only the tip of an iceberg. Remove the sweet feed and molasses from the diet and the problem will solve itself. The systemic yeast infection occurs in other breeds and species as well, but is most obvious in the LaMancha because the short ears allow the build-up of yeast to be easily seen as it drips from the ear.

CHAPTER 3

HERBIVORE SIMPLE STOMACH WITH HORSE AS EXAMPLE

Organs

See Ruminant Digestion for more details.

Esophagus

In the horse, the esophagus is approximately four to five feet in length. The esophagus ends in a sphincter that closes completely and tightly, unlike in other species, making it almost impossible for a horse to vomit or burp up gas unless the sphincter has become compromised (such as from an ulcer) (Getty, 2013, p. 10). Sheep and goats can vomit, although they do so rarely. Carnivores vomit semi-regularly, and this is a common and safe method for removing undigestible items from the body.

Salivary Glands

Horses can produce twelve quarts of saliva a day, stimulated by food present in the mouth. What the horse eats determines amount of saliva produced, roughage like hay causes much more saliva production than does grain. Saliva acts as an antacid in the digestive system. Too little saliva can lead to ulcers since the horse's stomach produces acid continuously, not just in the presence of foodstuff. To stay healthy, horses need to eat or have access to roughage food all the time (Getty, 2013, pp. 10-12). This is very important to remember, especially if horses are being fed concentrates to compensate for heavy work or stabling. Even horses that are in competition, work or have to be stabled (which is to be avoided if possible) need access to large amounts of roughage and, ideally, pasture.

Stomach

While described as simple, a horse's stomach is still somewhat divided, with a glandular area at the base and squamous region at the top. The glandular area is protected by mucus production while the squamous area has no protection from acid and can be prone to ulcers when digestion is disturbed (Getty, 2013, p. 10).

Compared to ruminants and to the rest of the horse digestive tract, the stomach is small, with a two to four gallon capacity. Unlike ruminants that can take in large quantities of forage but then rest and chew cud, horses need to be fed smaller amounts continuously since they are unable to take in large amounts at once (Getty, 2013, p. 11).

A horse's stomach produces HCl continuously, even when empty. This has some side effects when horses are used for work and recreation. If a horse is stabled or not fed before working, stomach acid can more easily slosh up into the unprotected squamous region, leading to ulcers. Allowing horses continuous access to feed mitigates this effect (Getty, 2013, p. 12).

HCl (Hydrochloric Acid)

Used as part of the immune system, stomach acid kills pathogens ingested with food, and activates pepsin, a digestive enzyme. Horses do not need a high pH like carnivores since they are not breaking down meat for protein. Because of this, horses fed antacids continuously may be unable to properly digest food and assimilate nutrients. Particularly problematic are the antacids that change how the stomach acid is produced, like H2 blockers and proton-pump inhibitors (Getty, 2013, p. 12).

After entering the stomach, food is churned and turned into chyme by HCl and motion. This chyme passes into the duodenum of the small intestine through the pyloric sphincter. From there it should not pass back into the stomach, but an empty stomach can allow chyme to return, damaging the pyloric sphincter and leading to duodenal and stomach ulcers (Getty, 2013, p. 13).

The above information illustrates the importance of simulating a natural feeding process for horses if continuous access to pasture is not available.

Small Intestine

Digestion in the small intestine of the horse is similar to that in ruminants and other mammals. Digestive enzymes such as lipase, peptidase, sucrase and maltase are secreted and help digest foods. Horses, unlike ruminants, cannot digest fiber in the small intestine because the fiber has not been broken down yet by fermentation (Getty, 2013, p. 14). See Ruminant Digestion for more information.

Cecum

The cecum in the horse, where fiber digestion takes place, contains billions of protozoans and bacteria, functioning similarly to a rumen but with much less space for food. Foods stay in the cecum of a horse about seven hours (Getty, 2013, p. 19).

Fiber from the small intestine enters the cecum at the top and descends to where it is mixed with the bacteria, water and digestive enzymes. Digested fibers move to the top of the cecum again to exit to the large intestine. In order for digested fiber to move upward, there must be a full cecum. Continuous eating promotes good digestion. As a horse eats, it takes in occasional dirt and sand from the ground and this eventually settles into the cecum. If there is not enough fiber in the cecum to push the sand out, the cecum will fill with sand and the horse will get "sand colic." Prevention includes allowing the horse access to food, fresh water and plenty of exercise, always (Getty, 2013, p. 19). In order to remove sand from the cecum, the horse needs to be fed a diet high in fiber. Addition of psyllium seed *Plantago ovata* (a mucilaginous herb that expands when soaked in water) to the diet can also increase bulk in the cecum and promote movement of fiber and sand from this organ.

Large intestine

The large intestine is where fermentation takes place and digestion of volatile fatty acids occurs (acetate, butyrate, propionate) (Getty, 2013, p. 20). Bacteria in the large intestine also produce heat, water and gas, although much less gas than ruminants. Sudden changes in feed can result in excess gas and improper digestion (Getty, 2013, p. 21).

Small Colon/Rectum

In the horse, there is a section of the colon that functions to reabsorb water and form manure before passing undigested material to the

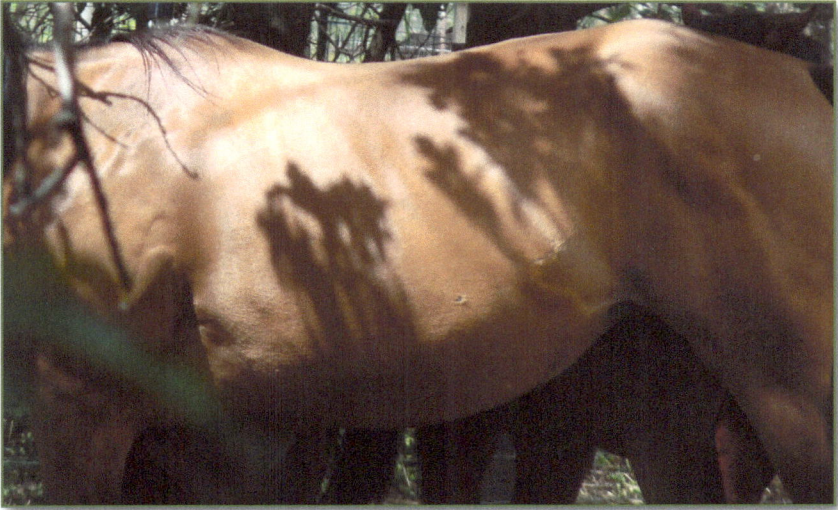

rectum and out of the body. Manure should not contain much undigested fiber or be excessively dry (indicating dehydration) or off-color. Changes in manure can indicate health problems (Getty, 2013, p. 21).

Natural Diet Considerations

Horses are designed to take in small amounts of plant material in a continuous fashion. They are also designed to move over large areas while eating and movement is as important to them as proper food and fresh water. Proper minerals are essential for health and these may not be provided in soils, plants or hay, and may need to be added as part of a supplement program.

See chapter on Minerals for more information.

Horses are not really adapted to eating high levels of carbohydrates, like grains, but if grain is to be a part of the diet, smaller amounts more often rather than a large feeding occasionally allows the digestive system to adapt to the grain and better process it.

Pastures with much diversity of plant species allow horses to eat what they need and keeps them from being forced to eat something poisonous because they have nothing else.

Pastures and Feeds

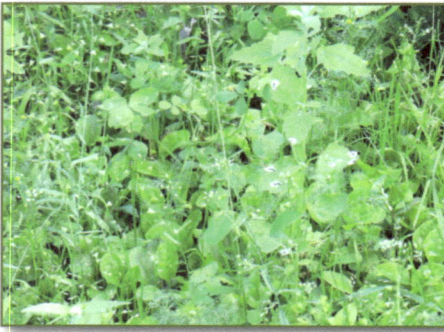

It is vital to true health for herbivores to have access to pasture. And the more varied the offerings, the more healthful. As was noted above, sheep learn from taste what is healthy and what is not. Feed concentrates and pellets do not allow them to taste and learn what is safe and nutritious. Lambs learn from their mothers what plants to choose and what to leave. Studies involving deer showed that deer select browse that is high in nutrients and leave growth from the same plant that is lower in nutrition. Animals can often self-medicate if medicinal plants are available for them to eat in pastures or as offerings.

Managing pasture can be difficult. Working with the environment instead of against it makes a farmer's job easier. Learn what plant species are native to an area and consider re-establishing these, since they are already adapted to local soil and

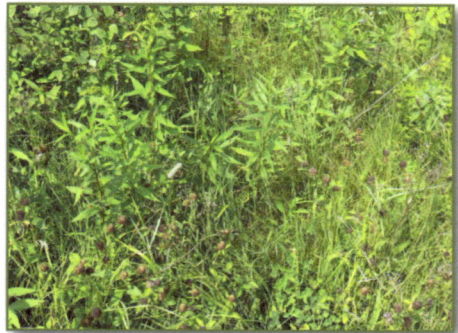

weather conditions, and will thrive where introduced species may not. There is much information available on the importance of Old World species for livestock, but native North American species have as many benefits, have the advantage of being locally adapted, and provide cover and food for wildlife, as well as for domestic livestock. This allows farmers to conserve diversity rather than destroy native plant and animal communities. Working with the land provides much benefit.

Many pasture plants are also medicinal in one form or another and animals may choose to self-medicate when allowed access to these plants.

CHAPTER 4

RABBITS

Natural Raising

Many rabbit breeders now feed only a pelleted commercial rabbit ration to their stock but this is absolutely not natural and contains artificial preservatives and flavorings, not to mention the extra treatment that must be done to pelletize a plant source. Vitamins and minerals must be added back in and products in pellets now may be GMO (genetically modified organisms).

Rabbits, like sheep, goats, horses and cattle, will choose to graze on pasture or browse along woods edges if given a choice. This is rarely practical because of the predators that will take rabbits. All types of raptors and owls, fox, coyote, dog and domestic cat, and all types of weasels will eat rabbit if available. Rabbit tractors (similar to chicken tractors) can be used for grazing or, more conveniently, rabbits can be kept in cages but fed appropriately for a foraging herbivore.

In summer, fresh pasture plantings can be fed, including clover (*Trifolium* spp.), alfalfa (*Medicago sativa*) (non-GMO!), dandelion (*Taraxacum officinale*), chicory (*Cichorium intybus*), garden vegetables (leafy greens and root crops), fruits, grasses and weedy species and tree leaves of non-poisonous varieties. Rabbits appreciate raspberry (*Rubus idaeus*), blackberry (*Rubus* spp.) and other bramble leaves, rose (*Rosa* spp.) leaves, as well as vetch (*Vicia* spp.) and other legumes like birdsfoot trefoil (*Lotus corniculatus*). Other plants and trees like aspen (*Populus* spp.), birch (*Betula* spp.), alder (*Alnus* spp.) and plant leaves like violet

(*Viola* spp.), plantain (*Plantago major* & *Plantago lanceolota*), chickweed (*Stellaria* spp.) and edible herbs are all favorite rabbit foods. Hay with a diversity of plant species is a welcome addition and can be fed or kept in front of rabbits continuously, and a grain mix can be used in small amounts. Flaked grains, oats, barley (watch for mycotoxins in barley especially), sunflower seeds, split peas, wheat, rye, etc. can all be mixed for a nutritious supplement. Kelp can be offered as a salt and trace mineral source. Kelp is not one species but a group of brown algae species in the Laminariales order. Fresh water should be available continuously.

It is important to remember that, if you purchased rabbits from breeders who have fed only pellets, you must switch the rabbit over slowly to a new diet of fresh feeds. It is not necessary to get pellets in order to make the switch. Instead, offer good quality grass mix hay as soon as your rabbits arrive. Give small quantities of fresh foods daily but always keep the hay available. As the rabbit starts to eat the fresh foods, you can increase the amounts over a two-week period until, after a few weeks, your rabbit can be eating a completely natural diet. If at any time the rabbit develops diarrhea, remove the fresh food, leave only good quality hay and plenty of water. If diarrhea continues, consult with a natural health practitioner versed in animal care.

CHAPTER 5

CARNIVORES

Mammals that hunt other animals for food are called carnivores. Most have four carnassial teeth adapted to shear through flesh (Macdonald, 1993, p. 18). Exceptions are the pandas that have teeth more adapted for grinding their vegetarian foods (Macdonald, 1993, p. 18). Modern carnivores, unlike their ancestors, are often more adapted to an omnivorous diet (Macdonald, 1993, p. 18).

Jaw muscle adaptation includes a strong temporalis muscle that allows for strong biting and masseter muscles for grinding and cutting food. Temporalis muscles allow for a strong bite when the mouth is wide open (for a killing blow) and the masseter muscles work when the mouth is closed (Macdonald, 1993, p. 21).

Example species: Wild species include lions, tigers, wolves, foxes and weasels of all kinds. Domestic species include dogs, cats and ferrets.

Simple Stomach Carnivore with Dog as Example

Ferrets and cats are sometimes referred to as obligate carnivores. This means they lack the ability to digest carbohydrates (vegetable or plant nutrients) and must get their nutrition from meat. Dogs are also carnivores but may have some adaptation to processing some carbohydrates. Wild carnivores that do eat some carbohydrates include fox species and coyotes. Under adverse conditions, many wild carnivore species may eat carbohydrates in an effort to survive. How much nutrition they gain from this is not known. Eastern coyotes, a hybrid amongst wolves, coyotes and domestic dogs, will eat everything from deer and small game to garden vegetables and garbage.

Body to Small Intestine Ratio in Carnivores:

Cat 1:4
Dog 1:6
Ferret 1:5.

Teeth:

The dental formula for a dog is 3142/3143 (total forty-two). Two incisors, one canine tooth, four premolars and two molars make up the upper teeth. Canine and incisor teeth are used for grasping, carnassial (upper fourth premolar and lower first molar) and other premolars are used for shearing and molars for grinding.

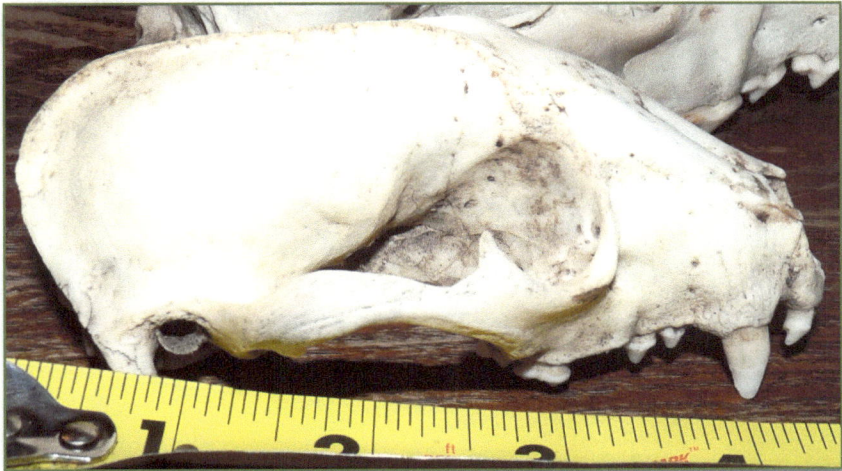

Dog breeds most close in form to wild canids (like the fox and wolf) have teeth that fit correctly into the jaw. Breeds that have heads flattened in front (brachycephalic) have teeth turned perpendicular to the jaw in order to fit into the mouth. In extreme cases of brachycephalic dogs, teeth may be missing and in other breeds, there may be extra incisors or premolars (Kainer & McCracken, 2003, pp. Plate 48–49).

Dogs' teeth are designed for grabbing rather than chewing. Their upper and lower teeth overlap rather than meeting when the jaws are closed. Dogs and their wild counterparts swallow much of their food barely chewed and almost whole.

Salivary Glands

Like other mammals, dogs have salivary glands that secrete saliva. In dogs, however, the saliva plays a role in cooling the body. Dogs do not sweat much, most of the cooling effect comes from airflow across the tongue that evaporates saliva. Dogs have parotid, zygomatic, mandibular and sublingual salivary glands (Kainer & McCracken, 2003, Plate 50).

Stomach

Food stays in the dog's stomach approximately six to sixteen hours where it is acted on by HCl and pepsin (Carlson, D. & Giffin, J. 1980, p. 169; Miller, et al, 1965, p. 681). The stomach is protected by mucus, as is the small intestine. Size varies by breed, from two cups to approximately two gallons (0.5 to 8 liters), and a 33 lb (15 kg) dog has an empty stomach weighing approximately 0.22 lbs (100 gm).

Small intestine

Food entering the duodenum is acted on by digestive enzymes amylase, lipase and others secreted by the mucus membrane (Kainer & McCracken, 2003, Plate 53).

Cecum

The cecum in dogs is small and not well-developed (Reece, 2009, p. 360). For more information, see below under Ferret.

Colon

The colon is where fermentation occurs in the dog.

Natural Diet Considerations for Dogs

Dogs, being carnivorous, have a digestive system designed to process a meat diet. Unlike cats and ferrets (see below), dogs do take in water by drinking, not just from food sources. Dry kibble, however, is still not a healthy choice for a dog for several reasons.

Protein sources in kibble may come from rendering plants and consist of diseased livestock that may have been heavily medicated. Some additives from rendering plants even contain euthanized dogs and cats! Toxic preservatives (like ethoxyquin) are used to keep the kibble from becoming rancid, grains are often added as filler and may be GMO (Zucker, 1999, pp. 12-13). The author of this book saw a national chain pet food manufacturer loading meat by-products from a rendering plant where companion animals and livestock were taken for processing after euthanasia! This practice is very real and very detrimental to the health of the animals being fed such horrors.

The diet of a dog may be as high as seventy-five percent protein. It is important to remember that wild canids would eat almost all of the animal they kill for food, including brains, intestines, bones, skin and fur or feathers. If it is not possible to feed whole prey to dogs, some careful balancing can be done to mimic a natural diet. In herbivores (the food source of carnivores), plant material is broken down and digested but still available to be eaten by the carnivore in the stomach and intestines. Plant material that is partially digested is similar in form to steamed or lightly cooked plant material. Feeding some cooked vegetables can be used in place of intestines, although this would lack the enzymes naturally found in raw food.

Raw bones are a natural part of the diet for dogs. As long as the bones are sized for each breed of dog and come from organic, healthy sources, they can be a healthy part of the diet. Long bones that are weight-bearing are more solid than bones like ribs and are more difficult for a dog to chew. If possible, avoid those weight-bearing bones or supervise feeding and remove any sharp parts the dog cannot chew. Do not feed cooked bones. These can splinter and lodge in the throat of the dog. Feeding any bones carries some risk but this is a risk that all carnivores take on a daily basis. If feeding a whole bone is not workable, always give calcium. Egg shells or powdered bone are good sources of calcium for dogs.

Carnivore Ferret as Example

Teeth

The dental formula for a ferret is 3131/3132. There are three incisors, one canine, three premolars and one molar on one side of the top jaw (Fox, 1988, p. 34).

Salivary Glands

There are five pairs of salivary glands in a ferret: parotid, mandibular, sublingual, buccal and zygomatic (Fox, 1988, p. 35).

Stomach

A ferret's stomach is similar to a dog's stomach. Shape and size vary depending on what and how much was eaten (Fox, 1988, p. 40).

Small Intestine

In ferrets, the ratio of body length to small intestine is 1:5 (compare to a dog's 1:6). The duodenum is short, the jejunum and ileum are not visibly distinguishable from the duodenum externally (Fox, 1988, p. 43).

Cecum

The ferret has no cecum.

Large intestine

The ferret does have a large intestine but it is difficult to separate from the small intestine although it is larger in diameter. The colon starts at the ileocolic junction, divided into three parts: ascending, transverse and descending (Fox, 1988, p. 45).

Liver/Pancreas

The liver is relatively large in ferrets, with the gallbladder extending through the lobes of the liver. The gallbladder never reaches the diaphragm like it does in a dog (Fox, 1988, p. 47).

Food can traverse the digestive system of an adult ferret in about three to four hours, less time in the younger animals (one hour in a newborn) (Lewington, 2002, p. 58)

Diet Considerations for Ferrets

Ferrets are carnivores, getting most of their nutrition from meat. They are sometimes referred to as obligate carnivores but this may be a misnomer as all carnivores get their nutrition from meat, fat and protein, not vegetarian sources. Unlike herbivores, they may have no carbohydrate requirement but get energy from fat (Fox, 1988, p. 141). Ferrets also may be unable to convert fructose (sugars from fruits) to glucose (the form of sugar used in the body of herbivores and omnivores) (Fox, 1988, p. 142). Unlike cats, however, ferrets can convert beta carotene to vitamin A (used for eye health, growth and muscle function) (Lewington, 2002, p. 59). Ferrets need ten amino acids from their food and will eat to try and satisfy this need, leading to a fat ferret with high protein in the urine (high blood urea nitrogen BUN). Taurine is one of the required amino acids in ferrets as it is in cats but is found in a meat diet (Lewington, 2002, p. 60).

Like cats, ferrets should and would take in liquid with their food if they were hunting and eating raw. A dry kibble does not provide water, and clean, fresh drinking water is essential. In fact, kibble diets may contribute to dehydration, kidney and bladder stones and bad teeth (Lewington, 2002, p. 63).

Commercial Diets

Several brands of kibble are available now for ferrets. Those that provide at least thirty percent protein and twenty percent fat from sources that are higher quality would be the best choice if you must use a kibble for part of their diet. Chicken meal contains more feathers and bone and less meat, so is deficient in amino acids and not recommended (Lewington, 2002, pp. 55, 58).

Pet and laboratory feeding schedules are often continuous but this does not mimic nature. A carnivore eats only when it is able to catch food and this is not continuous or regular (Lewington, 2002, p. 58). All carnivores are accustomed to some period of fasting when prey is not abundant or easily caught.

Carbohydrate-rich, sweet treats should be avoided for all ferrets and carnivores. These foods reduce essential fatty acid and protein intake and can lead to hair loss, weight loss and eventual death (Lewington, 2002, p. 58). Since ferrets cannot digest carbohydrates, kibble with less than two percent carbohydrates is best (or better yet, no carbohydrates at all!) (Lewington, 2002, p. 61).

Natural Diets

Since ferrets are carnivores, their wild relatives (polecats, weasels, skunks, etc.) would catch and eat several smaller rodent species and other small birds and mammals. A natural diet for ferrets might include mice, chicks, raw meat (like chicken wings) and an occasional raw egg. Raw eggs should not be a large part of any carnivore's diet (or

anyone's diet) since raw egg whites contain avidin (a glycoprotein). Avidin binds up biotin, one of the B vitamins and can lead to deficiency. Interestingly, raw egg yolk contains high amounts of biotin, which may help offset this deficiency if the whole egg is consumed. Nevertheless, including some raw egg as a small part of a natural diet is acceptable but continuous feeding of raw eggs should be avoided.

A simple raw diet for ferrets not fed whole prey (like mice) might include raw meat from several sources and some raw whole milk (not pasteurized or homogenized) to provide fat and calcium, plus fresh water (Lewington, 2002, p. 62). Any uneaten raw food should be removed to avoid spoiling.

Carnivore with Cat as Example

Cats may have originated in the desert and have some adaptations that allow them to survive well in a situation with little water. They require moisture from their food, however, and many cats have kidney problems when fed dry food because they will not drink enough water to meet their needs.

Teeth

The dental formula for cats is 3131/3121. Unlike dogs, cats have the upper premolar and lower molar shaped like carnassial teeth for cutting (Curtis, 2011, p. 2).

Tongue

A cat's tongue is covered with papillae that help remove meat from bones. Cats lack the ability to taste sweet, another indication their diet should not include carbohydrates (Curtis, 2011, p. 4). Some of the papillae contain taste buds and kittens have marginal papillae along the edges of their tongues (Kainer & McCracken, 2013, Plate 45).

Salivary Glands

Cats do not produce the enzyme amylase to start carbohydrate digestion but use hexokinase, an enzyme that does break down carbohydrates but is less efficient than the enzymes found in other animals for this purpose (amylase and glucokinase). Cats have parotid, sublingual, zygomatic and mandibular salivary glands.

Stomach

The cat's stomach is small compared to other species. Cats have some specific dietary needs compared to other species. They cannot make their vitamin D but must take this in as part of the diet. They also cannot convert beta carotene to vitamin A. They lack the ability to make the amino acid taurine from cysteine and methionine. All of these nutrients are found in a meat diet that includes eggs, meat, fish oil and animal fats. Water in food meets most of the metabolic needs of cats. Cats lack a strong thirst reflex and will not take in enough water if their diet is dry food (Curtis, 2011, p. 5).

Cats also lack the ability to convert linoleic acid to arachidonic acid, they must have arachidonic acid as part of their diet along with taurine and vitamins D and A (Curtis, 2011, p. 5).

In carnivores, vomiting can be a normal response to something indigestible in the stomach. Cats often vomit hairballs and other indigestibles, so this should not necessarily be seen as abnormal.

Small intestine

In cats, the small intestine is about three and a half times the length of the body (Kainer & McCracken, 2013, Plate 47).

Cecum

The cecum is non-functional in cats.

Raw Food Diets

The above information on carnivores is quite clear - a high meat diet is necessary for optimal health. But what about cooked meats? Canned foods and kibble have been cooked and processed, removing vital nutrients that are then added back to the foods. Nutrients like vitamins A and D and taurine that are found in raw foods but destroyed by excess heat must be added to commercial cat and ferret foods. From 1932-1942, Francis Pottenger did research on cats to determine amounts of adrenal extracts needed for health. His research provided unexpected information on cats' dietary needs. He maintained his original research animals on a laboratory formula of cooked meats but as the experiment progressed, he found these cats to be poor candidates for surgery, showing obvious signs of dietary deficiency. The research animal numbers began to exceed his supply of cooked meat and he resorted to sourcing raw meat scraps from the local meat packer. The raw meat-fed cats showed excellent health and reproductivity, while the cooked meat-fed animals continued to decline, even over generations, until they died of deficiency. Pottenger set up study groups to determine the result of the deficiencies in individuals and in populations of cats over generations. His results showed that cats fed raw diets had few parasites, allergies or other diseases, and reproduced easily with healthy litters of strong kittens. Those animals fed cooked diets had a myriad of health problems that extended to their kittens. Pottenger fed the kittens from deficient mothers a raw diet but found they did not completely recover the good health found in raw-fed animals. In fact, the poor health persisted for three to four generations even when the offspring were returned to a raw diet (Pottenger, 2012, pp. 1, 12-13)

Current understanding of cat nutrient requirements means commercial cat food manufacturers can add back in the nutrients destroyed by processing. But is this enough to restore and maintain true health in a carnivore?

Cats, like dogs and ferrets, can be fed raw or a modified diet that includes adequate protein and fat. Transitioning a cat or ferret to a new, raw diet can be a challenge. Patience and persistence help. Some websites now include information on raw feeding and making the switch to a different diet. http://feline-nutrition.org is one site that can be helpful.

CHAPTER 6

WATER

Animals can live without over half their weight in protein but losing a tenth their weight of water will kill them. Water cushions cells and makes cellular and tissue elasticity possible. The body contains about sixty percent water by weight (Reece, 2009, p. 28). It is needed to dissolve foods and excrete waste materials, and is a key component of blood and lymph; however, too much water will cause toxicity and death.

Most water comes from drinking, some from foods and some from breakdown of nutrients in the body (carbohydrates, fats, proteins). Some animals, like the kangaroo rat, can get all their water needs from the digestive breakdown of food. Most animals, however, need plenty of fresh water daily.

Water needs vary according to the animal (species, weight, age, whether the animal is pregnant, lactating or producing eggs), environment (temperature, humidity, air movement) and type of foods consumed. Water deprivation is more serious in a young animal than an old one. Water needs do not increase linearly by weight of the animal: for instance, a 1100 lb (500 kg) cow does not need ten times as much water as a 110 lb (50 kg) calf (Reece, 2009, p. 39).

Examples of differing needs by species include camel, sheep and donkey that can survive dehydration better than other species. The dromedary camel can survive a loss of twenty-five percent of body weight of water. It also can survive higher body temperatures during the day and allow the cooler night air to cool its body, meaning it does not need to lose water to keep cool (Reece, 2009, pp. 41-42).

As the temperature drops, the loss of water by evaporation from an animal's body decreases, a drop from 71.6 °F (22 °C) to 59 °F (15 °C), for example, reduces loss by half in steers. Forage from dry areas may contain only slightly more than half the water content of forage from wetter regions.

More fiber, minerals and protein in the diet increase the need for water. More protein requires water for dilution of urea, a by-product of protein digestion.

Some health issues arise from inadequate water intake, such as digestive complaints. Low water intake (such as from cold weather) can contribute to urinary calculi, or stones, in male sheep and goats (females also may have increased calculi but are able to pass the stones through the larger urethra).

Excess water can cause a reduction in roughage if feeds are fed soaked or fed as mashes (Abrams, 2000, p. 2).

Example Water Intake Needs

A lactating dairy cow (~1000 lbs/454 kg) needs about eight gallons (thirty liters) of water for maintenance and slightly more than a gallon per day for each gallon of milk produced. If water cannot be offered free choice, it should be given at least twice a day.

Working horses may need eight to ten gallons (thirty to thirty-eight liters) of water per day. A suckling sow taking in 16 lb (7 kg) per meal per day requires at least 50 lb (23 kg) of water.

Water to dry matter ratio of 3:1 is recommended (lbs of water: lbs of meal).

Sheep may need one gallon (3.8 L) per day but this depends on forage. Moderate rainfall can increase forage water content to eighty percent (ratio of 4:1, water to dry matter). High temperatures and low rainfall may create forage with 1:1 ratio.

Hens in egg production may need one-fifth gallon (0.8 L) per day of water.

Healthy animals will choose to drink the proper amount of water and fresh, clean water should be offered free choice to all animals. While it takes some energy to heat water to usable temperature once in the body, this expenditure is negligible. If the animal is being fed so poorly that the slight loss of energy is a problem, feed the animal better!

Excess water in stored feeds leads to mold (mycotoxin) growth and health problems in animals fed these contaminated feeds (Abrams, 2000, p. 5).

Not all water is the same. Pure water, not irradiated or treated, is the healthier option. Dehydration can cause an animal (or person) to not feel well enough to drink, in which case it becomes vital to get water into the animal. Without the appropriate electrolyte minerals, the body cannot keep water. Kidneys rely on potassium and sodium to either cycle water back to the body or eliminate excess water with wastes. Magnesium and calcium are also important for water balance and deficiencies in either of these can exacerbate dehydration symptoms in low water intake. By the same token, intake of these minerals requires water intake or toxicity can occur. Animals will die of salt poisoning if they have salt but no water.

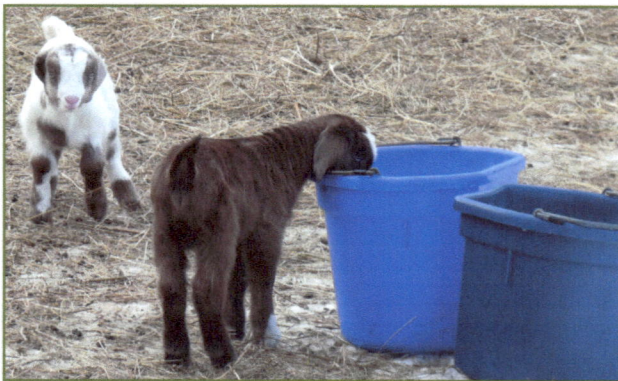

CHAPTER 7

VITAMINS AND MINERALS: THE BASICS

For carnivores, vitamins and minerals come from the meat sources they consume. Unless disease or digestive insufficiency occurs, a carnivore should not require supplementation; however, companion carnivores may benefit from a natural source of trace minerals, like certified organic kelp. Cod liver oil can provide a good source of vitamins A and D. Look for cod liver oil that is unflavored and has been filtered to remove the heavy metals that can concentrate in the oils of fish. Rampant pollution in our lakes, streams and oceans guarantees that fish and fish products will contain contaminants. Concentrated products like oils will contain high levels of heavy metals, PCBs and other pollutants if not filtered.

During times of stress and disease, carnivores and herbivores can benefit from the addition of other vitamins, like the B complex vitamins, vitamin E and vitamin C. More detailed information is included in the chapters on Vitamins and Minerals.

Herbivores are at the mercy of the plants available in their pasture to provide the needed vitamins and minerals. Plants, in turn, are reliant on these minerals being available in soils for their use. Different species of plants uptake minerals at different rates, and weather, soil pH and time of year all affect how minerals are used by plants and are therefore available to animals. Livestock are often fenced into pastures, not roaming hundreds or thousands of acres. This means the animals are only able to eat what is in that pasture, not seek needed minerals from other sources. In some areas of the country, for instance, in the past wild herbivores congregated to lick soils high in needed minerals. Areas of Kentucky are high in salt and minerals and town names reflect the history of animal use of high mineral soils.

Blue Lick, KY and Saltwell, KY are examples. Farms took names like Buffalo Trace Farm, reflecting their ecological heritage.

Modern farming methods can be detrimental to soil health and do not add back in necessary trace minerals to soils. Nitrogen, phosphorus and potassium (NPK) are used to boost plant growth, but excesses of these minerals can cause other trace minerals to be unavailable to plants, and therefore to grazing animals. Additions of these NPK and super phosphate fertilizers also do nothing to replenish trace minerals. Pastures where hay is sold off are very often depleted of minerals, the minerals being sold down the road with the hay.

Lack of topsoil, underlying rock structure, erosion and irrigation all affect mineral availability to plants and animals.

Having soils and forage tested gives information on what minerals are available to plants. Testing mineral usage in animals can be more invasive but gives a clearer picture of how animals are using the minerals available to them. Blood tests are not useful for determining levels of most minerals because circulating levels of many minerals depend on differing conditions in the animal. Liver analysis can provide much more information on mineral usage but depends on having a liver and that means a sacrificial animal. If the animals are kept for food, having a liver available is as easy as asking the butcher to save one. For a few minerals, hair sampling can be useful, but many labs do not do this or have accurate information on mineral levels.

Otherwise, mineral imbalances must be determined by looking at overall health of the animal, and toxicity and deficiency symptoms of the minerals. Supplemental minerals and salt are almost always offered to herbivores in some form. How and in what form the minerals are offered makes a huge difference in how available those minerals are for digestion and absorption in the animal.

Mineral Primer

Since sheep, goats, horses and cattle are herbivores, their nutritional needs are met mostly with hay and forage. Like all living things, they also need minerals, the building blocks of the body. They can take in much of their mineral needs from their feed, although almost every farmer offers some type of mineral supplement to their animals. In an ideal world, soils would provide most of the mineral needs for the animals through the pasture and hay, but many factors come into play in mineral availability in soils. Many farms have soils deficient in several key minerals, have applications of chemical fertilizers that cause trace minerals to be unavailable for use by plants (and therefore grazing animals) or have excessive pH causing certain minerals to become less available. Weather may cause plants to either uptake more of or less of certain key minerals.

Problems in health can occur when the mineral supplements do not contain enough of a particular needed mineral, contain it in a form that is not particularly available to the body for digestion and absorption, or don't contain the needed mineral at all. On some farms, excesses of certain minerals in soils, water or feed may throw off the balance of related minerals or cause elimination of other trace minerals. Supplemental mineral mixes may not contain enough of the needed mineral to offset those local conditions.

Minerals are generally divided into trace (or micro) and macro minerals. Trace minerals are those that are needed in very small amounts, but this is not to say they are not critical to health. In fact, they may be the limiting factor in life or death! Macro minerals, like calcium, magnesium and phosphorus, are needed in large amounts. They are also critical to health and life, without the proper amounts of these minerals in proper balance with each other, animals will eventually die. Thankfully, many mineral imbalances do not cause death, at least not immediately. Low-level deficiencies often lead to the above-mentioned problems and contribute to loss of production in flocks and herds. Those types of deficiencies cost a farmer a lot of time and money.

Increasing Needs

Situations that produce stress in a flock or herd can change mineral needs and upset digestive balance. Stress increases the need for minerals like selenium (and related vitamin E), magnesium and copper. Heat and humidity, travel, illness and parasites can all increase mineral needs (and may show mineral deficiencies and imbalances). Tetany from travel (stress), grass low in magnesium or lactation are symptoms of low magnesium, for instance. And white muscle disease that may occur in heat-stressed adults or show up as weak young and retained placentas at birth are signs of selenium deficiency. Most mineral mixes cannot compensate for these situations, and farmers need to be prepared to change their supplement strategies depending on weather, time of year and increases in stress.

Pregnancy and lactation increase need for minerals. If a pregnant animal is not getting what she needs for proper reproductive health before breeding, she will not produce as many young (or may not cycle at all). Males may produce inferior sperm. Weight loss is an issue and is related not only to mineral intake but to quality of feed. Once a female is bred, she needs enough minerals in proper ratio to meet not only her maintenance levels but to grow fetuses and prepare for birth and lactation. Her needs increase as the fetuses grow. After birth the need for some macro minerals, like calcium and magnesium, increase tremendously to meet the demands of lactation.

Young animals born deficient in some minerals may not thrive and may even die before reaching maturity. Selenium and copper deficiency can lead to weak lambs, kids, calves and foals (for different reasons: low selenium is associated with white muscle disease and weakness, while low copper is associated with poor nerve function and "swayback disease" in sheep and goats.) If young do survive (usually after much intervention from the farmer), they may grow poorly and can even die spontaneously from things like copper deficiency. Selenium also relates closely to immune function and fertility and acts as an antioxidant (important in protecting red blood cells) so is important during more than just birthing. Copper is important throughout life for immune health, ability of the body to use iron (and therefore make hemoglobin in the blood), wool production and quality hair growth, bone growth and reproductive health.

Minerals and Parasites

Many farmers consider parasites part of raising animals and attack them as the enemy. While parasites can certainly be detrimental (and in some cases, deadly), it is not reasonable to try and eliminate all parasites. Healthy animals will not maintain high levels of parasites, making the farmer's job much easier. Restoring and maintaining health is paramount to dealing with parasites, and minerals play a huge roll.

Copper is one of the first minerals to be considered when talking about parasites because of its use as copper oxide wire particles (COWP) to kill some parasites in sheep, goats and cattle. Copper is needed for much more than just killing parasites in the rumen, however, and copper oxide is not a particularly well-absorbed form.

It is likely that flocks needing the COWP have other underlying mineral imbalances contributing to higher parasite levels.

Selenium's role in immune health translates to importance in building up immunity to parasites and recovering from parasite infections.

Studies done on cobalt show that lambs getting adequate cobalt had lower levels of parasites and better weight gain than those lambs getting no cobalt but regular dewormings. These three are not the only minerals important to health but their lack can definitely increase susceptibility to parasite damage.

Disadvantage of Mineral Blocks

Sheep and goats have different needs than other herbivores and humans even though all mammals require essentially the same minerals for proper health. Because sheep don't have upper front teeth, they don't make good use of mineral blocks that require an animal to either stand and lick the block for long periods or bite off what is needed. Mineral blocks also have the disadvantage of containing salt. Since sheep self-limit salt intake, this keeps sheep from overeating the block but it also means they will not make good use of mineral or trace mineral blocks, even if they have severe deficiencies. Horses, on the other hand, will eat salt to taste, and mineral flavored with salt may cause horses to over-consume. Manufacturers compensate for this by not including high levels of minerals in blocks.

Mineral in blocks is also not in a very absorbable form. Blocks often contain the elemental forms of minerals that are not as easily digested. Mineral blocks and loose supplements contain high levels of salt and also contain flavor-enhancing additives and preservatives. The flavor enhancers ensure animals eat the mineral but they don't contain needed nutrients and can often be detrimental. Things like molasses add sweetness and may increase susceptibility of the animals to internal and external parasites and insects, and can disrupt proper digestive function.

Availability of Minerals

The form minerals are in is important. Elemental forms of minerals are not as easy for mammals to digest in many cases. Soil bacteria and fungi assist plants in taking the elemental minerals and changing the form to something useful for growth and much more digestible for animals. Mineral supplement manufacturers make use of this knowledge to create minerals that are bound to amino acids and therefore more available for digestion and absorption. These minerals are called chelated. Chelated minerals have several advantages, such as better availability for digestion and palatability. Because chelated minerals are so easily digested (sometimes close to 100% useable by the body), they can help offset situations where other minerals cause elimination or poor absorption of the needed mineral. This interplay occurs for all minerals and is a problem when one mineral becomes completely unavailable due to the extreme excess of another antagonistic mineral. Examples are molybdenum completely inhibiting copper, or iron interfering with selenium. Other mineral interactions become a problem if minerals are out of balance with one another. Excess salt will deplete potassium (or the reverse) and excess calcium not only causes elimination of trace minerals from the body, it throws off the balance with magnesium and phosphorus. Several of the more common trace minerals (and some macro minerals) are now available in a chelated form for supplementation, making it easier to get a proper amount of those minerals into an animal.

The Importance of Natural

It is surprising to hear about farmers who spend a lot of time giving injections, particularly when those injections are to try and meet mineral needs or compensate for vitamin defiencies. The BoSe (Bovine Selenium) injectable is used to treat selenium deficiency in pregnant animals (although this is extra label), prevent white muscle disease/weakness in newborns and treat low level signs of white muscle disease in adults. From my point of view, injecting minerals to compensate for the lack of available minerals in a supplement or soils and forage is not the best way to tackle the problem. Not only is this an unnatural way to get a mineral into an animal, it is costly, time consuming and can have side effects, in the extreme case death from anaphylactic shock. A better way to meet selenium needs would be to use a chelated product or work to improve soil to provide minerals to plants and animals. Chelated minerals have the advantage of crossing through the placenta, meaning young are born strong and running, crossing into milk, meaning young continue to get selenium from their mother after birth, and being bioavailable, meaning it is easily digested and there is no interference with antagonistic minerals. An ounce of prevention really is worth a pound of cure when working with minerals.

In addition to chelated mineral products, things like kelp contain needed minerals (both trace and macro) in an easily-digested form that is palatable. Herbs, brush and trees can be used to provide minerals in individual cases and as part of a grazing program. Since tree roots reach deep into the subsoil and bring up trace minerals otherwise not available in the pasture, animals may eat bark to try and meet trace mineral needs otherwise lacking in the supplementation program.

Improving soil health by reducing use of chemicals that can kill off beneficial bacteria, fungi and insects improves availability of minerals and nutrient content of forage. Restoring minerals through applications of compost and fertilizers (preferably not conventional) also improve mineral availability.

It is important to remember that everything needs to work together for animals to be healthy, and one of the first steps is proper digestion. Without this, animals cannot make use of any other ingested nutrient. All minerals interact with one another, in soils, in plants and in the body. While minerals are discussed separately, in reality, they cannot be so easily divided.

Minerals and Digestion

See Chapters 9 and 10 for more information on minerals.

The previous chapters on the digestive systems of herbivores provide information on how plant material is turned into available nutrients for health, growth, reproduction and healing. It should be clear that no matter how much quality feed is provided, if an animal is unable to digest it, the nutrients are worthless. Herbivores, and ruminants in particular, have very elaborate digestive systems designed to take plant material and use it to meet the nutritional needs of the animal. Ruminants can make use of roughage completely indigestible to other mammals, as long as their rumens are fully functional. Minerals play a huge role in maintaining the health of the bacteria in the rumen that are the basis for digestion and assimilation of nutrients.

Cobalt is a mineral I consider the base of the pyramid for ruminants. Although it is a trace mineral, it is vital to the rumen bacteria health and ability to assimilate nutrients. It is also used to make vitamin B12 (cobalamin) and is important in folate metabolism in all animals. Lack of cobalt results in rumen bacteria death and associated lack of appetite and weight loss (referred to as chronic wasting disease). Other symptoms include coldness, poor wool/hair production and/or loss and pernicious anemia, which is indistinguishable by sight from iron-deficient anemia. Iron-deficient anemia can be related either to low iron or low copper making iron unusable. Low cobalt also can cause tearing from the eyes. Sheep and goats can pull cobalt from stores in the liver to make vitamin B12 but they cannot use cobalt stored in the liver to put back into the rumen to support bacteria health. Cobalt must be taken in orally on a regular basis for sheep and goats to survive. Cobalt toxicity is much less a concern in ruminants than other mammals. Ruminants tolerate high levels compared to needs, although young who do not yet have fully functioning rumens should not be fed extremely high amounts of cobalt. Other herbivores can survive on pastures low in cobalt where ruminants would die. Cobalt is just as essential to horses but is not needed in as high amounts for proper bacteria growth and function in the digestive system (see below under Cobalt for more information).

A note about anemia, for those people fortunate enough not to have encountered it yet: mucus membranes should be a bright pinkish red, not pale pink or white. To find mucus membranes, you can check inside of mouth (gums), inside the vulva, or the easiest in sheep or goats, pull down the lower eyelid and check the tissue color beneath the eye.

Sulfur, a macro mineral needed for amino acid and protein production, is also critical to digestion. It is important for wool, hair, skin, hooves and nervous system health. Without proper amounts of sulfur, animals lose their appetite and have poor quality wool and hair growth. They may become susceptible to external parasites like lice, salivate and have tears from the eyes. Excess sulfur will interfere with trace minerals, reduce rumen and digestive function, and result in breath that smells like sulfur.

Many other minerals are important in digestion, including salt. Deficiencies and excesses will disrupt rumen bacteria and lead to poor health and possibly death.

Vitamins

Walk into any pharmacy or big discount store and you will see aisles of vitamin supplement products. The array is dizzying and leads one to believe we would never survive without some bottle or liquid product. In both animals and humans, there are certainly cases where supplementing vitamins in critical, such as when digestion is disrupted and the ability to make and absorb vitamins is diminished. In times of stress or illness, increasing need for vitamins makes supplementation a wise choice. Restoring digestive function is the first step, and reducing underlying causes of deficiencies should be next.

Herbivores make and absorb their B vitamins and vitamin C if they are healthy with properly functioning digestive systems. For those who keep guinea pigs, however, vitamin C is a concern and fresh fruits and vegetables should be offered to meet the vitamin needs. Otherwise, grazing should meet nutritional needs unless something changes substantially. During situations where digestion is poor (disease, parasites, stress), working on digestion while injecting vitamins is the quickest way to restore health. This is one of the only times I use injections: since the animal is not digesting properly, feeding the vitamins is a waste of time.

Vaccination is a stress and introduction of toxins, and increases the need for antioxidant vitamins like vitamin C. Oral (either ascorbate or herbal sources) or injectable vitamin C can be useful in mitigating some side effects of vaccination. See Volume 2 Vaccination chapter for more information, cautions and concerns.

Vitamins A, D and E are fat-soluble, meaning they are stored in the body, unlike vitamins B complex and C, which are easily flushed from the body and needed on a daily basis. Vitamin E, while fat soluble, is often considered more like a water soluble vitamin because it is not stored as well as A and D and may be needed on a fairly regular basis for health. Fortunately, forage usually meets the vitamin E needs of herbivores. Stored hay loses vitamins rather quickly, however, and supplementation may be necessary during periods of prolonged hay feeding.

Vitamin D is made by the body in the skin and sunlight is needed for proper amounts. During long winter months or prolonged periods of confinement in buildings, animals will need additional vitamin D. Vitamin D is critical for sheep to be able to use related minerals, like calcium, magnesium and the trace mineral boron. Vitamin A is generally adequate in forage but again, hay feeding means supplementation may be necessary.

For short-term oral supplementation of vitamins A and D, cod liver oil is a good source. Adult sheep and goats can be give thirty cc of cod liver oil daily for a short time to replenish levels. Lambs and kids can be given ten to twenty cc.

Human-grade vitamin gel caps for vitamin E can be used and B complex or C capsules or liquid can be fed. Herbs containing high amounts of these vitamins are also a good way to get quality, easily-digested forms of the vitamins into sheep. Rose hips (*Rosa* spp), dandelion leaves (*Taraxacum officinale*), parsley (*Petroselinum crispum*) and nettles (*Urtica* spp) are all good sources of several key vitamins and minerals. Keep in mind though that not all plants contain the same levels of vitamins or minerals, depending on the above-listed factors for uptake of minerals and production of vitamins and volatile oils. Different batches or species of rose will contain varying levels of vitamin C and how the rose hips were harvested and stored also affects amounts of vitamins available. Ideally, animals should be healthy enough to make their own B vitamins and a mineral supplement can provide the fat-soluble vitamins needed during winter.

Differing Needs by Species and Breed

Goats and sheep are quite similar in how their digestion works, and their mineral needs are also similar; however, there are differences in the amounts needed. Sheep need copper but there is a very fine line between enough and toxicity for sheep. Goats tolerate copper in higher levels. Some sheep breeds are more sensitive to copper, and some store copper more easily than others. Because of this, mineral companies generally market sheep mineral that contains low or no copper. This may work well for most sheep breeds in most situations but it will not work well for goats or some sheep on some farms. It is worth doing some soil, forage and blood/liver testing to determine what minerals may be needed in a supplement for a particular farm.

Horses and cattle need and can tolerate higher levels of minerals like copper, but even then some breeds of cattle need more copper and some do better with less. Whether this differing need is enough to change feeding and management is not something I will try to address here. See below under Copper for more information on breed differences.

General Comments

It is important to note that while vitamins and minerals are discussed individually, they are never used alone in the body. Vitamins, minerals and other nutrients form complexes for digestion and storage. Deficiency and toxicity symptoms may be similar for those nutrients that are related metabolically, and enough or excess of a related vitamin or mineral can in some cases compensate for the deficiency of a related nutrient. This does not mean the related nutrient should not be provided! See below under Selenium and Vitamin E for examples of these interactions. By the same token, extreme excess of one nutrient can throw off the balance of related nutrients and cause all types of metabolic problems in the animal. Vitamins and minerals need to be in balance with each other for true health.

CHAPTER 8

THE VITAMINS

Fat-Soluble Vitamins (A, D, E and K)

Vitamin A

Vitamin A is one of the fat-soluble vitamins that occurs in plants not as vitamin A, but as carotene (provitamin A). Herbivores convert carotene to vitamin A (a long chain, fat-soluble alcohol). Cats are unable to convert betacarotene to vitamin A, however. Fish oils contain vitamin A but it is slightly different, containing another double bond at the molecule level. It is only forty to fifty percent as biologically active as the vitamin A form retinol (the alcohol form) but it is a good choice for cats and other carnivores.

Carotenes are orange-yellow and occur in leafy greens and orange-yellow vegetables. There are many carotenoids but only a few are active biologically and therefore important in providing nutrition to animals. Alpha carotene, beta carotene, gamma carotene and cryptoxanthine (in corn) are important provitamins, with beta carotene being the most active.

Several conditions can cause changes in the vitamin A structure and reduce activity. Light, heat and moisture can affect vitamin A content and activity in hay. The digestibility of the carotene is affected by type of forage (including plant species and type of product, silage, hay, pasture, etc.) and harvesting technique (McDowell, 2000, pp. 18-19, 22).

In ruminants, vitamin A may not be well-absorbed or digested; however, the absorption rate may also differ with amounts available. When low levels are available, absorption increases and when toxic levels are available, absorption by the body decreases (McDowell,

2000, p. 22). The change from beta carotene to vitamin A occurs mostly in the intestines but there are species differences in absorption and storage. Animals that have the ability to absorb carotene have yellow fat and milk, this can be breed-specific. For instance, Holstein cows easily convert beta carotene, so do not store it in fat or milk. Their milk fat is white. Jersey and Guernsey cows do not convert beta carotene as well, storing it instead in fat and milk, so that their milk fat is yellow (McDowell, 2000, p. 23).

Because vitamin A is a fat-soluble vitamin, fat in the body is important for absorption of the vitamin as are other fat-soluble antioxidants (like vitamin E). Absorption also requires enzymes secreted by the pancreas and by bile salts. Some vitamin A can be converted to retinol and then to palmitate that is secreted into the lymph, while some is changed and moves into the portal blood. The lymph transports the vitamin A to the liver (bonded to a lipoprotein) where it can be stored or re-esterified (changed back into the vitamin A form) and transported by the blood to tissues (McDowell, 2000, pp. 24-25).

Vitamin A can be reabsorbed from the intestine and cycled through the liver, unneeded amounts are removed from the body with feces. The liver stores about 90% of the body's vitamin A, while the rest is in the kidneys (particularly in the dog and cat), adrenal glands, lungs, blood and other tissue (McDowell, 2000, p. 27).

Needed for growth and health, without vitamin A, animals go blind or can have night blindness. Deficiency also produces bone defects, reproductive problems (fetal reabsorption and problems producing sperm) and defects in growth. Abnormal growth in epithelial cells can result in keratinization (formation of hard substance) during deficiency. The keratinization can affect digestive, reproductive, pulmonary and urinary systems, making the animal much more susceptible to infection. Pneumonia and digestive problems, like diarrhea, are symptoms of vitamin A deficiency (McDowell, 2000, p. 30). Vision problems include dryness of the cornea, problems with dim or bright light vision and color vision, night blindness, ulceration of the cornea and severe lacrimation (tears from the eyes) (McDowell, 2000, pp. 32-33). Keratinization can result in bacteria and other pathogens more easily entering through the skin, lungs, digestive and urinary tract. Regeneration of tissue is inhibited and risk for disease and time to recover are increased. Risk for urinary stones is increased because damaged epithelial tissue interferes with normal urine flow. Cerebral spinal fluid pressure can be elevated in animals that are vitamin A deficient (McDowell, 2000, pp. 34-35). The glucosamine molecule requires vitamin A for production. Protein synthesis and RNA metabolism can be abnormal in deficiency (McDowell, 2000, p. 35). Placentas may be retained after birth and newborns can have pink eye, diarrhea, nasal discharge or seizures (McDowell, 2000, p. 51). Animals may reduce feed intake and therefore lose weight and become deficient in other vitamins and minerals (McDowell, 2000, p. 52).

Vitamin A deficiency reduces hatchability of eggs in birds and fertility in rabbits. Turkeys require more vitamin A than chickens and signs of deficiency in birds includes decreased egg production, emaciation and weakness along with eye ulcers, nasal and eye discharges, ataxia and death (McDowell, 2000, p. 57). Bone development depends on vitamin A. Without it, bone cells cannot be reabsorbed by the body and excess depositing of bone occurs. This can result in disorganized bone growth, irritation and inflammation of joints and muscle incoordination or other nervous system symptoms. A restriction (or decrease) of the area where auditory (hearing) and optic (eye) nerves pass can result in deafness and blindness (McDowell, 2000, p. 36).

The immune system relies on vitamin A, and deficiency will result in increased risk of disease (including mastitis and pink eye), increased time for healing and risk for reoccurrence and disruption of normal adrenal gland function.

Oxidation is a major cause of vitamin A destruction, heat without oxygen is less a concern. Hay cut during bloom stage or earlier and not rained on will contain more vitamin A than hay cut once plants have gone to seed or hay that has been rained on before baling. Hay left out to cure loses about half its vitamin A content in one day. Legumes have a higher percent of vitamin A than other plants but poor quality alfalfa may contain less vitamin A than grass (McDowell, 2000, p. 44). Stored hay (assuming it is stored out of sunlight, dry and not overheated) may lose seven percent of the vitamin A content per month stored.

 Pelletization is even more detrimental to vitamin A content, destroying up to forty percent of the vitamin A. Feeds and supplements containing vitamin A should not be stored for long periods (McDowell, 2000, p. 44-45, 48).

Zinc is needed for available vitamin A to be useable. Zinc deficiency in ruminants contributes to low vitamin A levels even if vitamin A is available in the diet (McDowell, 2000, p. 48).

Because of body stores (over ninety percent of which are in the liver), sheep and goats on vitamin A-deficient diets can do well for three to six months. Goats, being selective intermediate feeders, choose leaves that tend to be higher in vitamin A, giving them an advantage, unless they are forced to eat some type of commercial feed (McDowell, 2000, pp. 67, 164). Storage of vitamin A can provide enough vitamin A for a newborn female to complete her first pregnancy (McDowell, 2000, p. 164).

Some sources note that overhead power lines may deplete vitamin A stores (possibly all antioxidant vitamins and minerals) and this is a concern in more and more places. Animals grazed under power lines may require extra supplementation of vitamin A and other antioxidants (Coleby, 2010, pp. 40-42).

Blood testing is not necessarily an accurate representation of vitamin A levels since the body will pull vitamin A from the liver to the point of depletion in order to maintain blood levels. Liver analysis and cerebrospinal pressure can be used to evaluate vitamin A levels and testing the animal for night blindness is a good indicator of vitamin levels (McDowell, 2000, pp. 68-69).

Vitamin A is most easily included in mineral supplements, although it can be used as a liquid supplement in water or orally or as an injection. Vitamin A supplements for minerals are generally stabilized and include bioavailable forms of vitamin A (McDowell, 2000, pp. 69-70).

Several health issues can cause vitamin A deficiency in the body, including illness and aflatoxins (toxins produced by mold and fungus). Like the other antioxidant vitamins and minerals, vitamin A can be supplemented during times of stress and illness to meet increased needs of animals (McDowell, 2000, p. 70).

It is worth noting that while the above information tends to consider beta carotene and the other carotenoids as synonymous with vitamin A, in fact these carotenes have actions in the body in addition to their function in being converted to vitamin A. A natural diet and access to grazing a varied selection of plants insures that animals have access to the many antioxidant carotenoids in the plant world (McDowell, 2000, p. 75).

Vitamin D

Vitamin D in animals is cholecalciferol (D3) and in plants is called ergocalciferol (D2). In most animals, sunlight is needed for the body to make vitamin D3 from a precursor, and this reaction can take several days from first exposure to the UV light. Animals in confinement, fed hay for many months of the year or in areas of the world where part of the year has few hours of daylight will require supplementation to prevent deficiency (McDowell, 2000, p. 91). The skin is also responsible for making other metabolites of vitamin D, important in immune function and prevention of diseases (McDowell, 2000, p. 110). Vitamin D in products is at risk for oxidation and becoming rancid unless stabilized. Heat, moisture and oxygen exposure and trace minerals can all prevent vitamin D from being available to the body (McDowell, 2000, p. 94).

Vitamin D can be obtained from the diet and absorbed in the small intestine in the presence of bile salts, although this ingested form is only about fifty percent available. Most of the vitamin D needed to meet the body's requirements comes from sunlight (McDowell, 2000, p. 97). The exceptions to this are cats and dogs, which do not make much vitamin D in their skin. They lack the precursor 7-denydrocholesterol in the skin that is converted by the body to vitamin D when exposed to the UV sunlight (McDowell, 2000, p. 97). In chickens, the skin of the legs and feet is where vitamin D3 is made (McDowell, 2000, p. 97).

Like many vitamins and minerals, more is not necessarily better. Once the amount of vitamin D needed is met, the skin does not convert more precursor to D3, no matter how much longer it is exposed to the sun. Growth and production do not increase in response to extra vitamin D (McDowell, 2000, p. 98). Vitamin D is taken from the skin to the liver where it undergoes changes to a form of vitamin D that then circulates through the blood (McDowell, 2000, p. 99). Most animals do not store high amounts of vitamin D (although they can store higher amounts of vitamin A). Rather than actual storage of the vitamin, accumulations can occur in skin and fat tissues because turnover rates are lower in these tissues. The blood does contain high levels of the vitamin.

One of the metabolites of vitamin D is used in mineralization of bone and suppressing parathyroid hormone secretion. Parathyroid hormone is secreted by the parathyroid glands and is mainly used to increase concentration of calcium ions in the blood.

In animals, calcium transfers across the placenta to meet growing bone needs in the developing fetuses. Lambs can have enough vitamin D from this placental transfer to prevent rickets for the first 6 weeks of life and supplementation of pregnant female animals is a valuable way to provide for the vitamin D needs of offspring (McDowell, 2000, p. 102). Ruminants may adjust absorption of calcium from the intestine in reaction to needs by the body. Increased needs during periods of growth (like in young animals) results in increased absorption. Increased absorption and efficiency of this process increases during pregnancy and lactation (McDowell, 2000, p. 107). When dietary intake of calcium cannot meet the increased needs, calcium will be taken from bones. It is crucial to provide enough calcium during times of greater use (pregnancy and especially early lactation) (McDowell, 2000, p. 108).

Vitamin D and its metabolites are excreted in feces with the aid of bile salts.

Vitamin D is very hormone-like in function and the metabolite 1, 25-(OH)2D is a steroid hormone. Its main use is to maintain calcium and phosphorus ions in the plasma to support mineralization of bone and body function. It is used in RNA production and vitamins D and A deficiency are both linked to osteoporosis. As was noted above, parathyroid hormone and vitamin D both act to maintain calcium ions in the blood plasma to prevent tetany and regulate phosphorus levels (McDowell, 2000, pp. 103-104).

Calcium absorption from the intestine depends on vitamin D. During times of vitamin D deficiency, calcium absorption from the intestine can drop from almost fifty percent to as little as fifteen percent in humans (McDowell, 2000, p. 105). Vitamin D also plays a role in calcium and phosphorus reabsorption from the kidneys (McDowell, 2000, p. 107). See under Minerals for more information on phosphorus and calcium.

Deficiency

The effects of vitamin D on bone include not only mineralization but reabsorption. Deficiency leads to rickets in young animals and osteomalacia (softening of the bones) in adults. The range of organs that need vitamin D in some form include pancreas, bone, ovaries, brain, parathyroid glands, stomach, skin and mammary glands. The pancreas needs vitamin D for insulin production, and many types of cells require vitamin D for growth. One metabolite of vitamin D acts as an immune system hormone and aids in lymphocyte production and reduction in leukemia, colorectal and breast cancer cells. Deficiency of vitamin D is associated with prostate cancer and metabolites of vitamin D are sometimes used in psoriasis (McDowell, 2000, p. 110).

Vitamin D requirements are generally met by sunlight (especially at midday) in herbivores and humans, unless exposure to sunlight is lacking. Some conditions that may lower levels of UV exposure include confinement, weather conditions (fog, haze, smoke, pollution), season or northern latitudes (especially in winter when vitamin D production in the skin may cease), use of sunscreen and color of the skin or coat. Dark-skinned animals are less effective in making vitamin D in the skin. In carnivores, however, there is a nutritional requirement for vitamin D even if the animal is exposed to sunlight. A raw diet that includes some liver and fat will provide the vitamin D needed, but cod liver oil can be added for additional vitamins A and D (McDowell, 2000, pp. 111-112).

Need for vitamin D can increase when the ratio of calcium to phosphorus becomes larger than normal (see under Minerals below for more information) and the amounts of either mineral can change the need for vitamin D. Grain feeding can increase need for vitamin D, calcium and phosphorus because it raises the rumen pH. Potassium increases absorption of phosphorus but other minerals can interfere with phosphorus absorption, including aluminum and iron (McDowell, 2000, pp. 113-114). See under Minerals below for more information.

Vitamin D3 and D2 can be found in the diet. Alfalfa and red clover hay (sun-dried) can be high in vitamin D2, as can birdsfoot trefoil. Leaves on cut plants still produce the vitamin if they are in the sun. Vitamin D3 can be found in high amounts in cod liver oil and other fish liver oils, eggs and cow's milk (whole) in summer. Winter milk is very low in vitamin D3 (McDowell, 2000, p. 116).

Vitamin D deficiency results in rickets, a condition of low calcium and phosphorus. In young animals this presents as problems with cartilage, in older animals, with movement of the minerals out of the bones, this results in osteomalacia. Symptoms are essentially the same as those for calcium and phosphorus deficiency (see Minerals for more information). Which bones are most affected depends somewhat on the species of animal (McDowell, 2000, p. 118).

Other symptoms include anorexia, digestive problems, stiffness in walking, weakness, tetany and convulsions. Joints may be enlarged and breathing labored, legs bow, and young animals are born weak or deformed. Muscle tension can twist the bones in chronic deficiency. Excess bone and cartilage accumulate at ends of bones so that ribs may appear to have lumps and joints become deformed (McDowell, 2000, p. 120).

Young may be born deformed or weak if the ewe or doe was deficient during pregnancy, but if the mother had enough vitamin D stored or adequate sources in the diet, this should not be an issue (McDowell, 2000, pp. 120-121).

Vitamin D, in addition to calcium, is needed for proper milk production. Deficiency of either can result in milk fever (parturient paresis). In cows, milk fever occurs within three days of birth and more often occurs in older animals that do not respond as well to the hormonal form of vitamin D as younger animals. In sheep, it can occur before birth up to ten weeks after lambing. Changes in food, weather, stress or fasting can induce it, and in sheep, like in cattle, the result is collapse and eventual death (McDowell, 2000, pp. 121-122).

Symptoms of deficiency in birds manifests as rickets and hypocalcemia that results in stiffness and lameness with weak legs. Bones can become crooked and in broiler chicks, the deformities seen at a young age relate to low vitamin D and can be prevented by access to vitamin D (McDowell, 2000, pp. 125-126). Allowing birds access to fresh greens or pasture eliminates vitamin deficiencies that would otherwise develop on a grain-only diet (McDowell, 2000, p. 127).

For horses, vitamin D deficiency can cause not only weakness and deformities but also loss of appetite. Since sunlight is needed for vitamin D production, horses stabled for long periods need to be allowed access to the sun for a few hours around peak sunlight (noon). Or better yet, 24/7 pasture! (McDowell, 2000, p. 127).

Deficiency can be caused not only by a lack of the vitamin but also endocrine system malfunctions, digestive disorders, liver or kidney problems and medications. Heavy parasite loads may also inhibit vitamin D conversion due to possible liver or kidney damage and some grains, particularly rye and soy, can inhibit vitamin D absorption (McDowell, 2000, p. 139).

Toxicity

The fat-soluble vitamins can accumulate to toxic levels in the body (vitamins A and D being the most problematic). Excess vitamin D is associated with high calcium levels in the blood and tissues, leading to inflammation, calcification and cell breakdown. Organs affected include bones (decalcification occurs), joints, kidneys, lungs, parathyroid glands, pancreas, lymph and blood vessels and eyes. Animals exhibiting toxicity lose appetite and weight, have heart problems, muscle and joint problems and kidney failure (McDowell, 2000, p. 141).

High levels of calcium in the diet can increase vitamin D toxicity, just as low levels of calcium decrease toxicity. Addition of vitamin A to the diet decreases toxicity. How the vitamin is provided plays a large role in potential toxicity. Oral ingestion, even at very high levels, does not usually cause toxicity, while injections can (McDowell, 2000, p.142).

A note here about injections: while there are certainly times when injections of vitamins may be necessary to save an animal (or person!), injection is not a normal means of getting anything into the body and it runs the very real, increased risk of toxicity and introduction of other materials (preservatives, carriers and contamination). When at all possible, a natural source and method of getting a vitamin into the body is much preferred to injection.

Some plants can contribute to toxicity although there are not many found in North America. One introduction, yellow oat grass (*Trisetum flavescens*) can cause calcinosis (deposits of calcium in the tissues) in animals grazing it because it contains vitamin D3 (McDowell, 2000, p. 143). The cure is to remove the animals from the pasture where the plant occurs (or remove the plant from the feed). If done early in the toxicity, the animals should recover (McDowell, 2000, p. 143). In countries other than the U.S., these types of plant species that cause calcinosis can be of great concern to farmers and animal producers.

Vitamin E

Of the vitamins discussed so far, this one is probably the best known for its use as an antioxidant and partner to selenium. Vitamin E, although a fat-soluble vitamin, is eliminated from the body quickly and tends not to accumulate. Important as an antioxidant, for immune health, and for prevention of a myriad of diseases, it is critical in the diet and necessary year-round. Vitamin E is needed for the body to use selenium (see below under Minerals) and as such, deficiency leads to dietary muscular dystrophy and other selenium deficiency conditions.

Vitamin E in nature occurs in eight forms: alpha, beta, gamma and delta tocopherol, and alpha, beta, gamma and delta tocotrienols (McDowell, 2000, p. 257). The body absorbs and uses natural forms best. Rarely do supplements contain all eight forms, and usually not in forms that are necessarily best used by the body. Allowing animals access to pasture is the best way to meet their vitamin E needs. In supplements, studies have shown that the different carriers (emulsification, dry, liquid, etc.) are all used fairly well by the body, although synthetic forms of the vitamin (even those that involve starting with a natural form and then using methylation) are not (McDowell, 2000, pp. 158-159). There are d and dl alpha tocopherol forms, the d form is better absorbed. The carrier and esterification are important to bioavailability also, dl alpha tocopheryl acetate proves better than the succinate form for availability, and alcohol forms are better yet. Young animals, ruminants in particular, use the alcohol forms better than the acetate forms (McDowell, 2000, p. 159). Most mineral supplements contain vitamin E, however, it may not be in high enough levels or in forms best suited for absorption. Supplementation is never a substitute for proper feeding.

The tocopherol form, like the other vitamins, can be damaged by oxidation, and this effect is increased by heat, moisture, trace minerals (like iron or copper), alkaline environment and rancidity. Alpha tocopherol is an antioxidant that is destroyed during the process of reducing oxidation in the body. In order to further protect the vitamin, manufacturers esterify it, improving stability. This is why the dl-alpha-tocopheryl acetate and d-alpha-tocopheryl acetate are used (McDowell, 2000, p. 160).

Digestion of fats, including vitamin E, depends on pancreatic lipase and bile. Absorption of the vitamin is in the small intestine. Elimination of vitamin E from the body can be as high as eighty percent (in rabbits, chickens and people) and increased intake does not necessarily correlate to increased absorption. If the body has all that it needs, it will simply eliminate most of the excess. Esterified forms are less likely to be damaged in the digestive process than the natural tocopherol form, leaving more for antioxidant activity. Vitamin E is taken from the liver where it is transported in the LDL cholesterol (note that LDL cholesterol plays a beneficial role in the body).

Colostrum contains high levels of vitamin E, but the vitamin does not cross the placenta well at all. Newborn animals must have mother's colostrum or risk severe vitamin E deficiency (McDowell, 2000, p. 162). Deficiency affects the immunity and disease resistance of the newborn (McDowell, 2000, p. 163). Once newborn ruminants do ingest colostrum, however, their vitamin E levels are very high and continue to remain high as long as the mothers have adequate vitamin E available. Many farmers are now attempting to compensate for the unnatural process of either not making sure pregnant animals have adequate vitamin E during and after gestation, or removing newborns before they nurse or, worse, pasteurizing colostrum. Compensation techniques include injecting selenium and vitamin E (and sometimes even vitamins A and D) and giving some type of nutrition paste to the newborn at birth. This is a completely backward way of viewing health! Rather than supporting the natural process that gives a newborn what it needs to start life healthy, interventions and interference in a natural process (and interrupting bonding between mother and infant) is now the norm. This is not natural and, unless there is an emergency, should not be the normal routine when caring for pregnant and lactating females or newborns.

Although vitamin E is processed in the liver, it is stored throughout the body, in fatty tissues, muscle and liver (McDowell, 2000, p. 163). Excretion rates are high through bile, and storage does not provide long-term availability (McDowell, 2000, p. 164).

Vitamin E is necessary for many physiological functions, including in the muscular system, reproductive, immune and nervous systems (McDowell, 2000, p. 164). Interestingly, some of the basic roles that vitamin E plays in the body can also be filled by selenium and some of the amino acids, like cystine and methionine. There are most likely other important functions for which vitamin E is critical that are not yet known (McDowell, 2000, p. 164). Functioning as an antioxidant, vitamin E protects cells and tissue from free radical and oxidative damage that can be the result of environmental factors, proper immune response and basic cellular metabolism. Selenium is also an antioxidant, functioning similarly to vitamin E and in conjunction with it to protect fats and other cells from damage (McDowell, 2000, pp. 164-165). Vitamin E and selenium are also important in protecting against cell leakage of damaging by-products, like creatinine, and in helping protect red blood cells and capillary walls (McDowell, 2000, p. 166). The antioxidant nutrients (like vitamin E, selenium, vitamins C and A and carotenes) all interact. Deficiency in one can be somewhat offset by enough of the other nutrients, even to the point that vitamin E deficiency muscular dystrophy can be offset by the other antioxidants (McDowell, 2000, p. 166). Because vitamin E is so important in preventing oxidative damage from polyunsaturated fats, diets high in these fats often lead to vitamin E deficiency because the vitamin is used up during antioxidant activity.

Sperm integrity is maintained in part by vitamin E and platelet aggregation is limited due to the protective effect on arachidonic acid (McDowell, 2000, p. 166).

All the antioxidant nutrients are important in preventing tissue damage and diseases by supporting immunity. One of the important jobs of vitamin C is to regenerate the vitamin E to alpha tocopherol, in essence recycling the vitamin for use again. Selenium and vitamin E are needed to protect the immune cells that destroy invading bacteria and to help protect tissues from toxic by-products produced by the dying bacteria. This leads to depletion of these important nutrients during stress and disease. Vitamin E is also important in antibody production for protection against disease and in viral infections. Viruses that would otherwise not be of concern may cause disease in animals that have vitamin E deficiency. Supportive care for any ill or stressed animal should include enough antioxidant vitamins and minerals to protect and restore the immune system (McDowell, 2000, pp. 167-168).

In addition to the role vitamin E plays in immunity, it is important in the synthesis of DNA, in removing toxic metals from the body (in conjunction with selenium) and in protecting against damage from other toxins. This includes protecting muscles from damage due to antibiotic monensin use for coccidia in young of ruminants and pigs (McDowell, 2000, pp. 168-1690).

Selenium and vitamin E work together; selenium can spare vitamin E, while vitamin E and sulfur-containing amino acids can delay selenium deficiency. Deficiency in either nutrient results in tissue damage from oxidation.

Vitamin E is also important in other reactions in the body, such as ATP formation, creatine phosphate and vitamin C formation, even vitamin B12 metabolism. The ATP stands for adenosin triphosphate, the energy transportation in cells. Vitamin D metabolism may be inhibited if there is vitamin E deficiency.

Requirements are hard to figure since vitamin E needs are so closely related to the other vitamins and minerals. Polyunsaturated fatty acids (PUFA) increase vitamin E requirements, and oxidation in oils causing rancidity also deplete the body's stores of vitamin E. McDowell (2000) in "Vitamins in Animal and Human Nutrition" states that the vitamin E requirement for dogs is five times higher when there is high PUFA in the diet (p. 171). Excess cod liver oil increases vitamin E and possibly selenium needs (McDowell, 2000, pp. 170-172).

Deficiency

Vitamin E needs include not only growth and reproduction, but also immunity. Amounts that support the former are not necessarily high enough to also support proper immunity. McDowell (2000) notes in "Vitamins in Animal and Human Nutrition" that fifteen IU of vitamin E is enough to prevent muscle abnormalities while fifty IU prevents red blood cell destruction, but 200 IU is needed for proper immune function (p. 173). Anything that increases burden on the immune system increases need for vitamin E: stress, disease, even exercise.

Vitamin E and selenium interaction was noted above, but it is worth mentioning that enough vitamin E can reduce requirements for selenium, and selenium in adequate amounts will reduce the need for extra vitamin E. Because vitamin E is fat-soluble, the body does store it to some extent (not as much as vitamins A and D), and selenium is also stored, giving a buffer for times of stress and dietary deficiency (McDowell, 2000, pp. 172-174).

Vitamin E is listed in supplements but forms are usually not listed separately. The alpha tocopherol form is the one most often included, although the other forms are necessary in the diet. Vegetable oils contain varying amounts of the different forms of vitamin E, with corn oil containing four forms while cottonseed oil contains only two forms. Wheat germ oil contains four forms in high amounts and other grains contain four forms in smaller amounts (McDowell, 2000, p. 174). Eggs, liver and plants all contain vitamin E, while milk may or may not contain much vitamin E, and there can be extreme seasonal differences. Colostrum has high amounts of vitamin E (needed by newborns). Alfalfa is rich in vitamin E. Vegetable oils in pet foods tend to be low in vitamin E due to solvent extraction, and vitamin E is often added back to pet foods for this reason (McDowell, 2000, pp. 175-176).

Processing of feeds reduces vitamin E content. In pelleted feeds, both vitamins A and E are destroyed. Other processes that destroy vitamin E include heating, dehydration (including drying hay) and exposure to high moisture, grinding and addition of minerals like iron, copper, zinc and manganese (McDowell, 2000, p. 177).

Deficiency symptoms vary by species but muscle weakness and destruction (muscular dystrophy) occurs in all species with vitamin E deficiency (McDowell, 2000, p. 178).

Ruminants

In ruminants, vitamin E deficiency (and selenium deficiency) often shows up as white muscle disease. This can occur at birth, with weak or dead young, but may also occur as the lambs, goat kids or calves grow, occurring anywhere from one to four months of age. In deficiency at birth, newborns are often unable to stand or stand for long periods and may be too weak to nurse properly. If the symptoms are not too severe and the newborn can be fed via tube, immediate supplementation of vitamin E and selenium can reverse the symptoms over time. This can be difficult, however, since it requires that the farmer be in attendance at birth and make sure the newborn gets colostrum, is warm and gets supplementation (McDowell, 2000, p. 181). Prevention is much easier!

Older animals may have deficiency symptoms after stressful events and damage to the heart muscle can result in sudden death (this can also occur with copper deficiency). There can be stiffness in the legs and tremors and eventual paralysis (from weakness in the muscles). Death follows. The back may be arched in deficiency and muscles will be white when necropsied. There can be weight loss and lowered immunity resulting in illness and increase in parasites. The heart muscle is affected and the animal may stop eating (see below under Minerals, particularly specific minerals cobalt, sulfur and related vitamins and minerals for other causes of anorexia). Reproductive function can be affected and ewes may be sterile during selenium or vitamin E deficiency (and other mineral deficiencies). Vitamin E and selenium in proper levels increase immune response to viral and bacterial infections (including chlamydia), deficiency leads to susceptibility to disease (McDowell, 2000, pp. 183-184).

Heifers have been shown to have increased abortions and stillbirths and fewer eggs released during deficiency. Higher levels of vitamins C, D, E and A improve sperm characteristics, sperm count and sexual activity (McDowell, 2000, p. 186). Retained placenta is a sign of selenium and/or vitamin E deficiency and deficiency in either nutrient can result in metritis (inflammation of the uterus) in dairy does (McDowell, 2000, p. 187).

Pigs

Deficiency of vitamin E is so closely related to selenium that the two nutrients are considered together. Deficiency results in similar muscular dystrophy seen in ruminants and in pigs, toxic liver dystrophy. Deficiency was not as common before the 1970s. After confinement became common (with no access to green plants), low selenium levels in feeds contribute to vitamin E deficiency. Use of moldy feeds (mycotoxins) in confinement now inhibit vitamin E digestion and increase need for the vitamin. Protein supplements low in vitamin E contribute to widespread deficiency problems as well. Different breeds of pigs have differing needs for selenium and vitamin E, the commercial meat breeds require higher levels of selenium. Symptoms of deficiency (when they occur) include difficulty moving, jaundice, discoloration of peripheral body parts (like ears) and difficulty breathing. Sudden death can occur without other symptoms and fast-growing young may be the first affected. In pigs, the liver problems associated with deficiency are often the most pronounced. Other symptoms include those associated with decreased immunity, susceptibility to mastitis and other reproductive issues and even bad skin (McDowell, 2000, p. 189). Boars also experience reproductive issues, deficiency causes loss of sperm motility and abnormal sperm (McDowell, 2000, p. 191). Deficiency can also increase risk of death from supplemental iron injections for anemia, while increasing levels of both selenium and vitamin E protects against this shock (McDowell, 2000, p. 193). Young piglets may die at around eight weeks up to sixteen weeks if deficient and deficient animals are much more susceptible to effects of stress (such as castration or fighting) (McDowell, 2000).

Poultry

Poultry, like mammals, are also susceptible to muscular dystrophy from deficiency; however, chicks can exhibit oozing of fluids from the skin (exudative diathesis) and encephalomalacia, resulting in ataxia and legs that cycle. Turkeys and ducks have similar symptoms, and ducklings with vitamin E deficiency may even have crippled legs.

Supplemental vitamin E may save the bird but the damage to the legs may not reverse. Allowing birds access to pasture will do much to prevent these conditions. Egg hatchability is reduced in deficiency and young that do hatch may be weak (McDowell, 2000, pp. 194-196).

Equines

Muscular dystrophy from deficiency in newborn foals results in weakness, inability to nurse and problems in breathing and heart muscles. In foals, as well as calves and other newborns, tongue muscles may be affected and ability to suckle is inhibited. Because the newborns cannot suckle, their bellies are empty but this is not to be confused with lack of milk production in the mother. Older animals exhibit more typical muscle weakness, poor locomotion and eventual continuous laying down. Deficiency also contributes to poor immunity and susceptibility to diseases. Adequate vitamin E protects against neurodegenerative diseases as well (McDowell, 2000, pp. 197-198).

Dogs

Like other animals, dogs show some of the same deficiency symptoms, including muscle changes and weakness, changes in sperm, problems during gestation and weak or dead pups at birth. Destruction of red blood cells and degenerative eye changes can occur in deficiency, and fluid retention in the skin. Anorexia, depression and death can also occur, with changes in the kidneys from deficiency as well. Some skin conditions in dogs are associated with deficiency and poor immunity increases susceptibility to mange and other parasites and diseases (McDowell, 2000, pp. 199-200).

Cats

In cats, deficiency manifests a bit differently and can occur when the diet is high in unsaturated fatty acids, such as fish oils. Yellow fat results and cats exhibit fever, anorexia and deposits of fat under the skin that can be painful (McDowell, 2000, p. 200). The common misconception about cats is that, because they love fish, they should have only fish; however, a diet of fish exclusively will result in deficiency. A well-balanced and varied diet of raw meat and eggs, or allowing cats to hunt, will prevent deficiencies. Keep in mind, though, that in some areas of the country where habitat destruction has limited songbird numbers, cats can have a very detrimental impact on remaining populations. In such cases, it is better to provide a proper diet for the cats rather than risk decimating our remaining wildlife.

Rabbits

Unlike the other species considered, rabbits do not respond to selenium in correcting a vitamin E deficiency. Rabbits do exhibit the same muscular dystrophy that occurs in other species during deficiency (McDowell, 2000, p. 202). Feeding rabbits fresh greens and good quality hay can prevent deficiencies, and is a more natural diet than the pelleted feeds available commercially.

Low selenium in rabbits will result in lowered immunity and problems with the skin. Flaking and susceptibility to mites increases with deficiency. Offering a good quality loose mineral in a balanced formula, and kelp, can prevent many of the problems that plague rabbits, particularly angoras.

Prevention is much easier than treatment in deficiency and animals already badly affected may not recover. For young animals, it is better to make sure the dams get good nutrition during pregnancy and lactation than to treat every single newborn in order to prevent deficiency. If the dams do not get what they need during pregnancy, treating the newborns with vitamin E, and possibly selenium, may be the only option but proper management should make this unnecessary in most cases.

In adults, adequate vitamin E improves immunity and prevents a wide range of diseases.

Although this volume does not cover treatment, it is worth noting here that homeopathic remedies can be of assistance in helping the body absorb vitamin E and selenium more efficiently. See Volume 2 for more information about treatment options.

Toxicity

Vitamin E, even though it is fat-soluble, does not accumulate as quickly in the body and has less potential for toxicity. Massive doses (in excess of 5000 IU) increase anticoagulant properties when there is an underlying vitamin K deficiency, but supplemental vitamin K corrects this effect (McDowell, 2000, p. 216). In general, maximum levels for people and rats can be in the 1000-2000 IU range (McDowell, 2000, p. 216).

Vitamin K

Also called prothrombin factor, vitamin K is usually considered in its role in blood coagulation (clotting). It is produced by intestinal bacteria, but problems in the digestive tract will inhibit production and absorption of this (and other) vitamins. Some plants are antagonistic to vitamin K, like *Melilotus officinalis* (yellow sweet clover). This plant can be safely grazed by ruminants when it is fresh or properly dried, but if wilted or damaged, accumulates dicumarols that are antagonistic to vitamin K and therefore have a blood-thinning effect. The dicumarols are produced from common, safe and natural plant coumarins by molds that attack the wilted plant. The blood-thinning effect can be extreme and result in death if enough of the wilted plant is eaten (McDowell, 2000, pp. 227, 230).

In reality, vitamin K is not one substance but several, and is needed for more than just blood clotting. It is a fairly stable vitamin not as subject to breakdown as other vitamins. The synthetic vitamin is water-soluble rather than fat-soluble (McDowell, 2000, p. 230).

The forms of vitamin K available differ in their usefulness. Vitamin K1 or phylloquinone (the form extracted from plants), is a better option for blood coagulation properties than K3 or menadione, and should be used in cases where dicumarol or warfarin poisoning is suspected (McDowell, 2000, p. 231). Another form, menaquinone, is a result of fermentation with bacteria (this can occur in the intestine) (McDowell, 2000, p. 229).

Vitamin K is absorbed with pancreatic juice and bile salts (the result of fat digestion). Anything interfering with fat digestion will also interfere with vitamin K absorption and availability. Estrogen increases the absorption of fat and therefore vitamin K. Male animals are more at risk for deficiency because of this. The different forms of vitamin K are absorbed differently but which form is more available to the body, and less likely to be excreted, is in question. Phylloquinone and menadione are the two forms most used, phylloquinone appears to be better for maintaining levels of vitamin K in the body (McDowell, 2000, p. 233).

Vitamin K is stored in the liver and other tissues, but storage time is not long and the body depends on continuous synthesis rather than drawing on stores in times of increased need or deficiency. Excess and metabolites are excreted in the feces (McDowell, 2000, p. 234).

Prothrombin and three other clotting factor proteins are converted by vitamin K. In the case of an injury, thromboplastin is released, and with calcium and other factors, facilitates the formation of thrombin from prothrombin. Thrombin (an enzyme) causes fibrinogen to convert to fibrin and form a clot by catching red blood cells (McDowell, 2000, p. 235). Vitamin K also acts as a coenzyme to convert amino acid alpha-carboxyglutamic acid (Gla) to alpha-carboxyglutamate, part of the process that allows the procoagulant proteins (like prothrombin) to act (McDowell, 2000, pp. 235-236). In summation, vitamin K not only has an active role in anticoagulation but in the process of clotting and maintenance of bone matrix.

Vitamin K needs are met by synthesis in the intestines (by such bacteria as *E. coli*) and by diet. Ruminants do not require dietary vitamin K but can meet all their needs from the diet. Rabbits and rats reclaim vitamin K during coprophagy (ingesting their feces). Chickens on the other hand, do require dietary vitamin K because they do not synthesize it well (McDowell, 2000, pp. 239-240).

Excesses of other vitamins and minerals can throw off the balance of vitamin K. High levels of vitamin A and calcium will interfere with vitamin K requirements.

Vitamin K is found in leafy greens that produce chlorophyll, and many vegetables and liver contain high levels. Alfalfa has a good amount of vitamin K; however, grains are usually low in vitamin K (McDowell, 2000, pp. 241-242).

Deficiency

Hemorrhage and failure to clot blood that can lead to death are the main deficiency symptoms. These can be induced by either actual deficiency or antagonistic chemicals like dicumarol or warfarin (rat poison). Antibiotics and other chemicals that interrupt intestinal bacterial synthesis and absorption also cause deficiency. This is important to remember when considering using antibiotics. Any antibiotic use must be accompanied by probiotics and should only be done when absolutely necessary (McDowell, 2000, pp. 244-245). Other causes of deficiency include confinement of animals (no access to pasture) and mycotoxins, or molds, in feeds. Problems with mycotoxins can be especially prevalent in barley. Not allowing coprophagy in animals can lead to deficiency. For example, rabbits kept in wire cages where fecal pellets fall through are inhibited from normal coprophagy. Differing requirements by breed, which are important in some pig breeds in particular, also lead to deficiency (McDowell, 2000, p. 246).

Toxicity

Under normal conditions, vitamin K found in plants at very high levels is non-toxic. Synthetic vitamin K can cause toxicity and result in several symptoms, including kidney failure (McDowell, 2000, p. 258).

Water-Soluble Vitamins (B Vitamins and Vitamin C)

B Complex Vitamins

B vitamins are often considered together because they are found together in foods and used together for coenzymatic reactions in the body. They are not stored in the body, excess of these vitamins is eliminated rapidly, and toxicity is much less an issue than with fat-soluble vitamins. The B complex vitamins are useful for nervous system health and restoration, and can be used either prophylactically to prevent problems during times of stress or illness, or as part of a treatment plan for nervous system damage. They should be considered part of a supportive care treatment plan and any time digestive function is disrupted. In ruminants, when digestion is not working well, oral supplementation of any vitamin is going to be ineffective. Conditions that can disrupt digestion include mineral deficiencies, antibiotic use, mycotoxins, stress or illness. In the case of the B vitamins, thiamine deficiency can lead to polioencephalomalacia and death. Deficiency with digestive disruption is one time when injection becomes necessary.

Toxicity is not an issue unless B vitamins are supplemented in high amounts for long term, or one particular B vitamin is taken out of balance with the others and interferes with the other B vitamins. Deficiency is much more an issue. Symptoms will be considered below under each vitamin.

The importance of a healthy digestive system and proper diet are illustrated again by the range of problems associated with B vitamin deficiencies. These, like all other nutritional problems, are much easier to prevent than to treat once a situation arises.

B1 (Thiamine)

Thiamine is found in whole grains, alfalfa, yeasts, and in small amounts in milk, liver and other organs, eggs and other animal protein sources. It is also found in small amounts in some plants and herbs, like rose hips (*Rosa canina* and *R.* spp.). Ruminants meet their thiamine requirements through synthesis by rumen bacteria, although anything that upsets the digestive system risks inducing thiamine deficiency in ruminants.

Thiamine is generally stable at low heat (100 °F), but like all the B vitamins, is soluble in water; therefore, moisture contributes to degradation by heat. Alkaline conditions also destroy the vitamin.

(McDowell, 2000, pp. 265-267).

Several substances act as antagonists to thiamine, including some thiaminase-containing plants (like *Equisetum* scouring rush). Thiaminase is an enzyme that destroys thiamine. Bracken (*Pteridium* spp.) is another example of a plant that can interfere with thiamine, usually in horses. In the case of thiaminase-containing plants, these plants are often completely safe when processed either by drying, heat or alcoholic extraction.

Coccidiostats, used to control coccidia in young animals, also inhibit thiamine. Other methods of control or prevention of coccidia can be used that are safe and do not interfere with thiamine or digestive function.

Other substances in some plants that can interfere with thiamine include the polyphenols, like tannic and caffeic acids. Endophyte-infected fescue also interferes with thiamine, and high levels of sulfur can cause thiamine deficiency, although rumen synthesis of thiamine is not hindered by sulfur/sulfate (McDowell, 2000, pp. 267-268).

Many fish species also contain thiaminase that is released after death as the fish begin to rot. Diets of raw fish in carnivores can lead to thiamine deficiency, although heating will negate this effect and animals that catch fresh fish are not affected.

(McDowell, 2000, pp. 267-269).

Thiamine is absorbed primarily in the small intestine from food sources but the digestion must include enough HCl. In some animals, the digestive bacteria manufacture thiamine. This is usually only absorbable if the animals are coprophages, or in the horse where thiamine is absorbed in the cecum (further down the digestive system from the small intestine) (see Herbivore Simple Stomach). In ruminants, some thiamine is available in the rumen (McDowell, 2000, pp. 269-270).

Thiamine absorption depends on the corticosteroid hormones and is passed easily through the placenta (McDowell, 2000, p. 270). The need for corticosteroid hormones in thiamine absorption underlines the importance of the glands in overall health, and the adrenal glands in particular. The Hypothalamic/Pituitary/Adrenal axis (HPA axis) controls so many body functions: immunity, vitamin and mineral absorption and other critical reactions. Stress, illness, medications and toxins that disrupt this system contribute to almost all diseases.

Like other water-soluble vitamins, thiamine is not stored in the body long, up to two weeks in mammals. Excess thiamine is excreted rapidly, except in pigs where stores can last up to two months. Thiamine occurs in higher levels in areas and organs of the body with the most activity (heart, kidneys, brain, liver), and is excreted in feces and urine (McDowell, 2000, pp. 270-271).

The Krebs cycle (citric acid cycle) in the body and mitochondria use thiamine, along with other B vitamins (niacin, riboflavin and pantothenic acid), in part to convert carbohydrates to energy that the body can use. One of the phosphate forms of thiamine also acts as a coenzyme in glucose breakdown in the liver, adrenal cortex, brain and kidneys. Thiamine forms are also needed for the production of ribose in RNA (ribonucleic acid) and DNA (deoxyribonucleic acid). Although the exact mechanism is not understood yet, thiamine is important in nerve function and nerve transmission, and in insulin synthesis (McDowell, 2000, pp. 273-274).

Requirements for thiamine have not been established for different species, and the ability of some species to make thiamine and absorb it make dietary requirements even more difficult to determine. In young ruminants (lambs, calves and kids) before their rumens are functioning, thiamine deficiency is of particular concern and can lead to polioencephalomalacia and death. Dietary intake plays a large role in meeting thiamine needs, low protein diets in lactating animals can lead to low thiamine, as can low carbohydrate diets. Body fat and protein can spare thiamine during times of deficiency and diets higher in carbohydrates increase thiamine requirement. Mycotoxins, antagonistic plants or foods (like some raw fish), high sulfur, diseases, stress and other factors increase need for thiamine. Anything that upsets the body's ability to digest runs the risk of thiamine deficiency. Requirements are also related to size of the animal and the amount of energy the animal is using (this can include hyperthyroid states, feverish conditions, young animals growing rapidly and lactation). Older animals do not make as efficient use of the vitamin, and their needs increase as ability to digest decreases (McDowell, 2000, p. 276). Heat and freezing can destroy thiamine in foods (including milk) (McDowell, 2000, p. 279).

Deficiency symptoms are similar in all species and involve the central nervous system. Beriberi occurs in humans, polyneuritis in birds and weakness in the legs is the start of thiamine deficiency in young ruminants. In ruminants, eventually the head swings back over the shoulders in a typical "star gazing" pose and the animal develops diarrhea and twitching eyes and dies. Adult ruminants can develop polioencephalomalacia. Polioencephalomalacia is actually a disease that causes brain degeneration, but the term is used to refer to the above symptoms of nervous system damage that respond to thiamine injection. There can be convulsions, blindness and circling before death in acute cases. Chronic cases may only show diarrhea and weight loss over time. This is also associated with cobalt deficiency and may be closely related, since cobalt is needed for proper rumen function and bacterial health (McDowell, 2000, pp. 283-284).

Some factors that can contribute to polioencephalomalacia and thiamine deficiency are high sulfur and gypsum use for regulating feed intake. In both sulfur and gypsum, sulfite is one of the forms during breakdown and can bind with thiamine, forming a thiaminase-type molecule that inhibits thiamine. High sulfur also interferes with copper and can form complexes with molybdenum and copper causing complete elimination and deficiency of copper (see Minerals below) (McDowell, 2000, pp. 285-286).

High-grain diets, large amounts of molasses and stresses (like gestation, lactation or growth) can also lead to polioencephalomalacia (McDowell, 2000, p. 287). McDowell reports in "Vitamins in Animal and Human Nutrition" (2000) that cattle can show extreme wasting, even with good-quality feed, that is alleviated by thiamine injections. He also notes that the condition can be prevented by proper mineral supplementation (p. 288).

In pigs, thiamine deficiency manifests not only as weight loss but with vomiting, colder-than-normal body temperature, nervous system problems and death from heart failure (McDowell, 2000, p. 289). Low-level deficiency is the same as for other species: weight loss and anorexia. Chickens also show weight loss and neuromuscular problems that include the twisting of the head and neck (like "star gazing" in ruminants) (McDowell, 2000, p. 290).

Although horses should not have problems meeting their thiamine requirements if their diet is good and digestion functioning well, in cases of deficiency their symptoms are similar to other animals and include reproductive failure (McDowell, 2000, p. 292).

Carnivores, like cats or dogs, have similar symptoms but causes can differ and include sulfur dioxide-preserved meats and raw fish diets (McDowell, 2000, p. 293).

When diets are good and there are no problems with digestion or thiaminase, thiamine deficiency should not occur. Most diets for livestock would be sufficient to meet or exceed requirements. Deficiency should be a sign that something is not right with the feed, the animals or the supplemental minerals. Mineral supplements that contain too little of needed minerals or too much of some minerals can be problematic.

For severe deficiencies requiring supplementation, oral supplementation is not usually sufficient because the digestive system is already impaired and absorption decreased. Injectable thiamine can be given, recommended doses for lambs or calves can be 100-400 mg/day (divided doses are best since thiamine is eliminated quickly from the body). Adult sheep could be given 500-2000 mg/day (goats would be similar) (McDowell, 2000, p. 303).

Toxicity is generally not a problem but injections of 100 times the recommended amounts can produce symptoms of allergic reaction and shock (McDowell, 2000, p. 305).

B2 (Riboflavin)

Riboflavin, like thiamine, acts as a coenzyme. This yellowish-orange vitamin occurs in every animal or plant cell and is made in the rumens of adult ruminants. Young ruminants, before their rumens are fully functional, must have a dietary source, although as long as they are getting milk, their needs should be met (McDowell, 2000, p. 311, 326).

Stability varies by pH, it is not destroyed by heat in neutral or acid conditions and is not particularly soluble in water. Water solutions are unstable in light, particularly in heat or alkaline solutions

In nature, riboflavin exists as three forms, including free riboflavin. Another form is a coenzyme derivative called flavin mononucleotide (FMN). Flavin mononucleotide is also called riboflavin 5 phosphate, produced when riboflavin combines with a molecule of ATP. The third form of riboflavin is flavin adenine dinucleotide (FAD), produced when FMN is combined with another molecule of ATP. Digestion frees the riboflavin from the latter two forms and absorption occurs in the small intestine (McDowell, 2000, pp. 312, 315). Specific riboflavin-binding proteins are responsible for moving riboflavin to the fetus in pregnant animals and may be controlled by estrogens. Thyroid hormone also facilitates conversion of riboflavin from the coenzyme forms and is under the control of adrenocorticotropic hormone and aldosterone (McDowell, 2000, p. 316).

Like the other water-soluble vitamins, riboflavin is not stored for long in the body, although liver, heart and kidneys do contain higher levels and retinal tissues also contain riboflavin. Urine is the main way the body excretes excess riboflavin, although some is also removed in bile, through the feces (McDowell, 2000, p. 317).

Riboflavin helps produce energy at a cellular level (ATP) and is also involved in the hydrogen transport system. It is important in carbohydrate, fat and amino acid metabolism. Riboflavin also functions to convert vitamin B6 to its coenzyme form. Deficiency of riboflavin affects vitamin B6 status. Mineral use in the body can depend on riboflavin. Iron needs riboflavin for absorption. Deficiency in riboflavin leads to iron deficiency and iron-deficiency anemia (McDowell, 2000, pp. 317, 320, 321).

Requirements, like those of most other vitamins and minerals, change with stage of growth, stress factors, environment, and amount of fat, sugar and protein in the diet. These factors limit body production, increase requirements and ability of the animal to make its own riboflavin. Generally, older animals need less; however, reproduction increases need for riboflavin (McDowell, 2000, p. 321). Amount needed depends on what species but varies from about 1 mg/kg of dry matter diet to 6 mg/kg dry matter diet (McDowell, 2000, p. 322).

Riboflavin is available in leafy greens (including alfalfa) and yeasts, but is low in grains (and removal of the bran removes almost all the riboflavin grains contain). Fermentation increases availability and animal products tend to be high in riboflavin (organ and muscle meats, milk, eggs). Milk contains more riboflavin in ruminants than is found in the diet of the lactating female because rumen synthesis of the vitamin is used in milk production (McDowell, 2000, pp. 324-325).

Light and heat destroy the vitamin but meat sources tend to be stable even when cooked (McDowell, 2000, pp. 325-326).

Ruminants and Rabbits

Synthesis of riboflavin in the rumen means adult ruminants should have no deficiency issues. This ability depends on proper rumen function, conditions like acidosis can inhibit synthesis or absorption. Sores and reddened mouth, anorexia, diarrhea, loss of hair and lacrimation or salivation can indicate deficiency (McDowell, 2000, pp. 326-327). Rabbits also appear not to have deficiency as long as digestion is normal and the rabbit is allowed to practice coprophagy (McDowell, 2000, p. 335).

Lambs, kids and calves have a dietary requirement that can easily be met if they are nursing. Animals raised on bottle, however, may not get the necessary riboflavin (in addition to a long list of other vitamins and minerals not available in milk replacers). When replacer is necessary, consider using a real milk recipe instead of a powdered replacer. See Chapter 11 for recipe.

Pigs

High-grain diets contribute to deficiency. Symptoms of deficiency include anorexia, hair loss, poor movement and stiffness, diarrhea, vomiting, light sensitivity, cataracts and other eye conditions, nervous system and skin conditions. There can also be reproductive problems, reabsorption of late-term fetuses, large front legs in piglets at birth or hairless piglets. Reproductive problems can occur before other, more obvious signs of deficiency are seen (McDowell, 2000, p. 327).

Poultry

Problems with the nerves in chicks lead to paralysis and curling of the toes in low level deficiency. Extreme deficiency leads to death without symptoms (McDowell, 2000, p. 330). Myelin sheaths of nerve cells become damaged and may not be reversible, although it is worth noting that homeopathic Hypericum (St. John's wort) has shown ability to aid the body in regeneration of nerves. Consider treating affected birds with both riboflavin (and all the B vitamins) and homeopathy for improved absorption. Prevention is much preferable and birds allowed access to pasture will not have this problem. Those confined for whatever reason should be fed a varied diet that includes fruits and vegetables, in addition to seeds and grains. A good natural vitamin and mineral supplement can be offered free choice also. Symptoms in chicks mirror those of other animals: anorexia, weakness and diarrhea. Egg production is reduced as is hatchability. Chicks hatched from hens that were deficient can exhibit deficiency and eventual death (McDowell, 2000, pp. 331-332).

Dogs

Again, symptoms are similar to other animals with deficiency and include reduced growth, lesions on the eyes, dermatitis, weakness, ataxia and death (McDowell, 2000, p. 333).

Cats

Cats exhibit similar symptoms, anorexia, hair loss, cataracts, incomplete development of testicles, fatty liver and death (McDowell, 2000, p. 333).

Final Note

Deficiency can easily be prevented with proper diet, access to pasture for herbivore species now raised in confinement and attention to proper digestive system health. Any supplementation should be in addition to all the B vitamins, since excess of one may interfere with absorption of the others.

Toxicity

Oral doses do not produce toxicity, even at very high levels (possibly as high as 100 times the requirement). Injectable doses can cause toxicity but the dose would need to be exceptionally high (McDowell, 2000, p. 342).

Vitamin B3 (Niacin)

Niacin is available as both nicotinic acid and nicotinamide, although the body can also create niacin from the amino acid tryptophan. Deficiency may relate to both the vitamin and the amino acid. Similar to the other B vitamins discussed, niacin functions as a coenzyme in systems where carbohydrates, fats and protein are broken down for energy in the body. The vitamin is not stored well, but some is available from the liver. Excess is excreted primarily from the urine. Niacin is relatively stable compared to other vitamins. Heat, light, air and alkalinity do not affect it; however, the nicotinic acid will form compounds of salts when exposed to calcium, copper, sodium or aluminum. The pH of the medium controls much of the interaction and degradation of the vitamin.

The ability of the body to convert tryptophan to niacin depends on several steps and nutrients. The body needs riboflavin, vitamin B6, protein, magnesium, iron and copper for the conversion to niacin. Deficiency of niacin can depend entirely, or in part, on the ability to metabolize tryptophan.

(McDowell, 2000, pp. 347-354).

Niacin is needed in the body for hydrogen transfer, which involves the coenzyme forms of nicotinamide, nicotinamide adenine dinucleotide (NAD) and nicotinamide adenine dinucleotide phosphate (NADP). It is also important in carbohydrate metabolism (the Krebs cycle and glycolysis), protein and fat metabolism, photosynthesis (in plants) and rhodopsin synthesis. Rhodopsin synthesis is part of the cycle by which the retinol form of vitamin A metabolizes in the eye to transmit energy to the brain that becomes vision. The NAD and NADP forms are used to synthesis and breakdown amino acids, like lysine, arginine and asparagine, and to repair broken DNA strands. Deficiency leads to damage of the DNA and problems with glucose tolerance (important as a precursor to diabetes mellitus) (McDowell, 2000, pp. 356-357).

Requirements for niacin are difficult to define exactly since the ability of the animal to convert tryptophan to niacin is an important factor in the amounts needed. This conversion relies on the availability of the related vitamins needed for the reaction, riboflavin and vitamin B6. It should be noted that the body will use free tryptophan for making proteins before making niacin. Feeds low in tryptophan will not provide extra tryptophan needed for niacin production. Other factors affecting niacin requirements are similar to those that affect the needs of B vitamins in general, including age and gender of the animal, stress (like disease), mycotoxins in feed, access to pasture, age of young at weaning, diet (different carbohydrates require differing amounts of niacin for proper use in the body) and species. Younger animals at weaning will need supplementation, making later weaning advantageous. An example of species differences include cats, which cannot convert tryptophan to niacin. Ruminants may not have a dietary niacin requirement because of their ability to make the vitamin in the rumen; however, things that can affect the digestive system in ruminants will change their need for supplementation. Balance of amino acids in feeds, energy content in the feed (higher energy feeds require more niacin), and antibiotics and other medications that potentially disrupt rumen function all affect need for niacin (McDowell, 2000, pp. 358-359).

In nature, niacin is not particularly rare, but is often in a form that is not available for digestion and assimilation. Yeasts, leafy greens and pasture contain available niacin but dairy, grains (especially corn), eggs and fruits are not good sources. For people, coffee and tea can be good sources of niacin, especially coffee from roasted beans (McDowell, 2000, pp. 60-362).

Deficiency

Deficiency symptoms include those related to the digestive system (anorexia, weakness, diarrhea, poor growth) and skin conditions. Diets that are low in niacin are not problematic to young ruminants as long as tryptophan is available. Diets low in both result in animals that become dehydrated, anorexic and weak, with diarrhea and then death. Adults appear to be able to synthesize niacin in the rumen to meet needs unless something interferes with proper digestion. The exception may be dairy animals that are bred to produce large amounts of milk. These animals may not be able to meet their own needs and require dietary supplementation. In times of stress and illness, supplemental niacin (and the other B vitamins, plus supportive minerals) can be important to health.

Niacin can be important in other conditions, like ketosis. In "Vitamins in Animal and Human Nutrition" (2000), McDowell cites a study finding treatment of dairy cattle with ketosis with niacin helped reverse symptoms (p. 364). Supplemental niacin increases the ability of rumen bacteria to make proteins, increasing weight, milk production and digestion. Ketosis can also occur subclinically, and result in weight loss, lowered milk production and poor feed conversion. Adequate niacin can improve all of those symptoms (McDowell, 2000, pp. 362-365).

In lambs, niacin helps improve protein digestion and assimilation. Most pigs do not have problems with deficiency, although this may vary somewhat with breed and diet. When deficiency does occur, it affects skin, digestive system and nervous system. Chickens and turkeys do not generally have niacin deficiency but again, this can depend on diet. Young birds have plenty of tryptophan in the yolk that is available from absorption but if the diet during growth is deficient in tryptophan, niacin deficiency can occur. The tibiotarsal joint swells, legs can become deformed, feathers don't come in and there is a lack of appetite and decreased growth in niacin-deficient birds. Niacin deficiency also causes black tongue, a swelling of the mouth and tongue with weight loss and poor egg quality, poor egg hatchability and death. Deficiency in other fowl is similar. (McDowell, 2000, pp. 367-368).

Horses do not appear to be susceptible to deficiency since they can manufacture niacin in the gut and from tryptophan in tissues.

Dogs, however, have deficiency symptoms similar to humans. These include black tongue (see fowl note above), anorexia, slowed growth, drooling of thick saliva, sores in the mouth that may be necrotic and can continue down the digestive tract, producing diarrhea and hemorrhagic necrosis of the small intestines. The skin is affected and irritation will cause the dog to scratch and bite, producing sores. Eventually, deficiency will produce death, often within 10 days of symptom onset.

Cats cannot synthesis niacin from tryptophan and deficiency symptoms are similar to those seen in dogs (McDowell, 2000, pp. 368-369).

Rabbits generally produce the niacin necessary for health in the intestinal tract when the diet is not adequate, but deficiency, when it does occur, is similar to other animals (McDowell, 2000, p. 372).

Because much of the feed available to grazing animals has niacin in a bound form unavailable for digestion, supplementation may be necessary for times when diet is insufficient, the animals' digestion is compromised, there is increased stress or illness, or during times of high growth rate. Nicotinamide and nicotinic acid are both available commercially and are well-used by animals. These can be fed and toxicity is not much of a concern when the vitamin is given orally. Amounts would need to be 10-20 times normal. Injection or IV doses can produce toxicity, but again, not until larger amounts are reached (McDowell, 2000, pp. 379-380).

Vitamin B6 (Pyroxidine, Pyridoxamine, Pyridoxal)

Vitamin B6 is not one nutrient but several related compounds that are important in enzyme reactions for fat, protein and carbohydrate metabolism. The plant form is pyridoxine, the common forms in animals are pyridoxal and pyridoxamine, although there are also coenzyme forms of pyridoxal phosphate and pyridoxamine phosphate (McDowell, 2000, pp. 385, 386).

Heat, light and an alkali medium are all destructive to the vitamin and some substances and drugs used in human treatment can interfere with, or bind with, vitamin B6. Linseed (flaxseed) oil contains an antagonistic substance and some pesticides can be antagonistic. In humans, contraceptives may also be antagonistic and result in deficiency. MSG (monosodium glutamate) sensitivity may respond to vitamin B6 supplementation, leading to the obvious conclusion that the MSG is antagonistic to B6. One of the reasons B vitamins are often given in complexes rather than as single vitamins is the potential for one vitamin to interfere with the absorption of another. Niacin in high doses can cause vitamin B6 deficiency. Any time one of the B vitamins must be given in higher amounts, the other B vitamins should also be given in recommended dosage (McDowell, 2000, pp. 387-388).

Absorption of the vitamin occurs in the small intestine (particularly the jejunum and ileum), even though it can be synthesized in the large intestine. In pigs, the cecum may also be a site of absorption. The dietary forms must be split from the protein molecules they are bound to and are then absorbed and transported to the liver. The PLP form (pyridoxal phosphate) is the form most active in metabolism, although all the forms are able to be changed from one to another. This conversion back and forth requires niacin and riboflavin (another reason to give B vitamins in complex). The pyridoxal form is the one that crosses the placenta, but alcohol inhibits this crossing (McDowell, 2000, pp. 389-390).

Many tissues contain vitamin B6 but muscle tissue contains most of the body stores, along with the kidneys. Excess is broken down and mostly removed by the kidneys, through the urine (McDowell, 2000, p. 390).

Vitamin B6 (mostly as PLP) is involved in energy production from the citric acid cycle (Krebs), carbohydrate, amino acid and fatty acid metabolism (McDowell, 2000, p. 391). The vitamin acts as a coenzyme for reactions that convert amino acids to histamine, serotonin, taurine and other amines. These reactions occur in all organs of the body. Deficiency of B6 in the brain can cause neurological problems (convulsions, etc.). Deficiency, or problems in passing the vitamin from mother to fetus, can result in fetal and newborn learning problems and neurodevelopment. Other reactions that involve vitamin B6 include metabolism of sulfur-containing amino acids, breakdown of glycogen to glucose, conversion of tryptophan to niacin and synthesis of epinephrine and norepinephrine. Use of iron in hemoglobin, antibody formation and amino acid transportation also involve vitamin B6 (McDowell, 2000, pp. 392-393).

The dietary requirements for vitamin B6, like the other B vitamins, depend on several factors: age, gestation/lactation, species, illness, digestive function, breed differences, and weather. In weather extremes, such as high temperatures leading to heat stress, vitamin B6 needs increase.

Ruminants may have no dietary requirement; however, young animals whose rumens are not yet fully functional and older animals with compromised digestion (for whatever reason) are going to need a dietary source or supplementation. Higher protein diets increase the need for vitamin B6 and high levels of thiamine cause deficiency of vitamin B6 (remember when giving injectable thiamine to treat polioencephalomalacia to also give injectable vitamins B complex) (McDowell, 2000, pp. 395-396).

Natural sources of the vitamin include muscle meat and liver, whole grains (especially in the bran), vegetables, yeasts and royal jelly from bees. How the feeds are stored affects the vitamin content; processing, cooking and storage can remove up to seventy percent of the vitamin B6 content. Oxidation, heat or light can also degrade the vitamin and irradiation can also destroy the vitamin (McDowell, 2000, pp. 397-398).

Deficiency

Deficiency of vitamin B6 affects immunity, including lymphocytes and tumor growth. Diseases like HIV and rheumatoid arthritis can show immune system changes that are associated with deficiency (McDowell, 2000, p. 394). Slowed growth, convulsions, anemia, and skin conditions (including hair loss) and lowered protein metabolism and tryptophan conversion are all associated with B6 deficiency. As noted above, adult ruminants generally do not have deficiency because they synthesize the vitamin; however, stressed animals with poorly functioning digestive systems may have problems and young animals on milk replacer can easily become deficient. Loss of appetite, loss of coordination (and demyelination of nerve cells), diarrhea (from breakdown of the intestinal mucus cells) and depression are signs of deficiency in ruminants (McDowell, 2000, p. 399). Deficiency in other species is similar and includes anemia and nervous system problems.

Some concerns exist for animals due to extreme lifestyles, like those with a higher protein diet that may increase need for vitamin B6. This is not the same as the higher protein that is normal in a carnivore diet. In dogs and cats, young that are deficient may have slowed growth, anemia and sudden death, although there will be more pronounced symptoms at necropsy. Rabbits show scaly skin around the ears and nose, inflammation and skin conditions on the paws, eyes and nose and neurological signs (including convulsions). The hind legs may become completely paralyzed (McDowell, 2000, pp. 403-405). Note that rabbits can occasionally break their backs by catching a leg in a cage or on rough handling; it is worth examining the animal and the diet to determine the cause of paralysis.

Supplementation should not be necessary in most livestock, not even in young animals as long as they are nursing. Simple herbivores and birds may need supplementation, depending on diet and other factors (stress, feeds, etc.) since not all feed contains adequate B6 and improper storage can cause degradation. Heat, light and moisture can affect levels in vitamin supplements. Mineral mixes with vitamins may have degradation if minerals are in the oxide or carbonate forms (McDowell, 2000, pp. 410-411).

Toxicity

Toxicity is rare since vitamin B6, like other B vitamins, is water-soluble and easily eliminated from the body. Excessive feeding of high levels, however, can lead to problems with the nervous system, convulsions and death. Spinal cord damage and testicular damage can occur with high doses and pyridoxal is the most toxic of the forms (McDowell, 2000, p. 412).

Pantothenic Acid

Pantothenic acid, like the other B vitamins, is found in enzymes (coenzyme A and acyl carrier protein) important in carbohydrate, fat and protein metabolism. Most all feed contains pantothenic acid, but not in high levels. Ruminants synthesize this vitamin but other species are at risk for deficiency (McDowell, 2000, p. 419). Some nutrients can be antagonistic to pantothenic acid. For instance, high levels of copper in chick diets will result in lower levels of coenzyme A formation (McDowell, 2000, p. 421).

Pantothenic acid is a yellow oily substance that is water- and ethyl acetate-soluble. Heat and extremes of pH destroy the vitamin, although the commercial form of calcium pantothenate is fairly stable.

Food sources of the vitamin are either free (as pantothenic acid) or bound coenzymes that need to be broken apart during digestion. Pantothenic acid is absorbed in the jejunum of the small intestine as is the alcohol form, panthenol (which is absorbed more quickly) (McDowell, 2000, pp. 421-423). The supplemental form, sodium pantothenate, is well-absorbed in dogs. Pantothenic acid is needed along with ATP (adenosine triphosphate) and cysteine to synthesize coenzyme A and is the limiting factor in this synthesis.

Storage is not high but can occur in liver and kidneys. Excess is eliminated in urine and also as a breakdown product of carbon dioxide from the lungs (McDowell, 2000, pp. 423-424).

As was noted above, pantothenic acid is part of coenzyme A and acyl carrier protein (ACP). Found in all tissues, coenzyme A is involved in at least 100 reactions in the body important for protein, carbohydrate and fat metabolism. Coenzyme A is also involved in the Krebs (citric acid) cycle and is important in synthesis of cholesterol, sterols and fatty acids and their oxidation. Acetate is the active form of coenzyme A, that as acetic acid forms acetylcholine that is important in detoxifying drugs in the body. Acetic acid can also be part of the Krebs cycle or form ketones. Ketones are substances created when the body breaks down fat for energy, as is the result when insulin is in short supply (McDowell, 2000, pp. 424-425).

Pantothenic acid is also important in the immune system, where it stimulates antibodies. Deficiency leads to low titers and decreased resistance to disease (McDowell, 2000, p. 426).

Requirements differ by species, breed and by stage of growth or reproductive cycle. Substances like antibiotics can decrease the need for pantothenic acid. In ruminants, as long as the digestive system is functioning well, there is no dietary requirement for pantothenic acid. What is in the diet also determines how well ruminants can produce the vitamin. Diets high in cellulose require more pantothenic acid than those higher in more soluble carbohydrates. Other species of simple herbivores and carnivores do make some pantothenic acid but absorption is assumed to be low unless the animals are coprophages (like rabbits).

There can be an interaction between pantothenic acid and other vitamins, particularly vitamins C, folacin, B12 and biotin. Diet can change how much pantothenic acid is needed, since the vitamin is so important in fat, protein and carbohydrate metabolism. Deficiency of pantothenic acid causes weight loss even when diets are high in fat. High protein diets can help compensate for weight loss from low pantothenic acid, this may be due to the need for pantothenic acid in coenzyme A carbohydrate metabolism (McDowell, 2000, pp. 427-428).

Many foods contain pantothenic acid, including alfalfa, molasses, yeast, green plants, organ meats, eggs, meat, milk, fruits and grains and royal jelly (a great source). While at first glance it is tempting to assume then that animals will have no problem meeting their needs through diet, commercially-raised animals like pigs may have deficiency issues (McDowell, 2000, p. 429). The vitamin is fairly stable to most manipulations, except those involving moisture, so dried or pelleted feeds are actually decent sources, as are those stored well (without moisture). Freezing and canning can cause large losses of the vitamin (most important for human food considerations) (McDowell, 2000, p. 430).

Deficiency

Effects of deficiency are similar in all animals although their actual symptoms may differ: skin conditions, nervous system problems, digestive problems that lead to poor growth and poor use of feed, immune problems related to lack of antibody formation and increased susceptibility to diseases and adrenal gland dysfunction.

Ruminants

Deficiency is due more to digestive impairment or interference, since the vitamin is synthesized in the rumen. When deficiency does occur, symptoms include scaly skin around the eyes and nose, other types of dermatitis, anorexia and digestive problems (diarrhea), weakness and eventual convulsions (from nerve damage). Susceptibility to infection (like pneumonia) and eventual death can also result from deficiency. Supplemental calcium pantothenate can reverse the damage.

Pigs

Like ruminants with pantothenic acid deficiency, pigs show skin conditions, loss of appetite, weakness and a particular "goose-stepping" gait in the hind legs due to nerve degeneration. They may also fall or have hind legs splayed to support themselves. There can be brown lacrimation around the eyes and all types of breakdown in skin (including in the digestive tract where lesions lead to bloody diarrhea). Adrenal glands and heart are affected and fatty liver disease may result. Infertility can result during deficiency, estrus is normal but embryos either do not implant or fetuses die before birth. Those piglets that are born to deficient sows may have poor suckling and impaired tongue movements (McDowell, 2000, pp. 430-432).

Chickens

In chickens the adrenal glands, skin and nerves are affected, and egg production and hatchability are decreased. Chicks hatched deficient will be weak and die. Those that survive will stop growing, feathers will not come in or become rough, and deficient birds will develop skin conditions, including on the feet, where infection may set in. Symptoms are similar to biotin deficiency (see below). Turkeys show similar deficiency symptoms as well, although ducks generally only have a loss of feathering. Like deficient pigs, there can be fatty livers and dilated hearts (McDowell, 2000, p. 435).

Horses

There has not been a deficiency established for horses and they can meet their needs through synthesis in the intestines (McDowell, 2000, p. 436).

Carnivores

Dogs and cats show similar symptoms to other animals, including anorexia, poor growth, lowered immunity and antibody production, weakness, diarrhea, vomiting and death. Vomiting can be very severe. Nervous system damage results in hind limb stiffness, convulsions and coma (McDowell, 2000, p. 436).

Supplementation should be considered for animals fed primarily corn, especially if they are not ruminants. Corn does contain bioavailable pantothenic acid, but the total is not as high as in other grains, like wheat. Forms available for supplementation include *d* and *dl*-calcium pantothenic acid. The *d* form is the active form available for use by the animal (McDowell, 2000, pp. 439-440).

Toxicity

Pantothenic acid is generally safe in higher doses, although higher-than-needed amounts are not particularly beneficial unless offsetting deficiency or problems with digestion and absorption (McDowell, 2000, p. 441).

Biotin

Biotin deficiency is not generally a problem in animals raised with access to pasture. Carnivores fed properly should also not have an issue with deficiency. Diets high in raw eggs, however, can lead to deficiency in carnivores. This can be offset by feeding liver and kidney, or possibly by feeding the whole egg raw since the yolk is high in biotin. Newer agricultural confinement methods for poultry and pigs lead to deficiency as well. Deficiency in animals results in skin lesions and hair loss (McDowell, 2000, pp. 445–446).

Like thiamine, biotin contains a sulfur atom in its structure and is fairly stable although it is soluble in hot water. Acids and alkaline substances, rancid fat, UV radiation and formaldehyde destroy the vitamin (McDowell, 2000, p. 446).

Biotin is found in bound and free forms in nature, the free forms being the only ones available for digestion by animals. In vegetables, fruits and milk, biotin is not bound but in meat, yeasts and seeds, biotin is bound with protein. An animal's ability to break that protein link determines whether the biotin is absorbable or not. The first part of the small intestine is the site of biotin absorption in most animals; however, pigs absorb biotin from the hindgut as well. Digestive bacteria create biotin but the biotin may not be available for absorption if it is synthesized past the part of the intestine where absorption takes place (McDowell, 2000, p. 449).

Biotin is in all cells, liver and kidneys and is transported by the blood. It is moved into cells from plasma by active transport and into the liver by a sodium transport system. Because biotin is synthesized in the intestine and often past the point where it can be absorbed by the body, excretion in urine and feces can exceed the amounts consumed. The kidneys recycle biotin for use again in the body (McDowell, 2000, p. 450).

Similar to other B vitamins, biotin is important in fat, carbohydrate and protein metabolism, and in conversion of these substances from one to another. Blood glucose levels are also maintained in part by biotin, and carboxylase enzyme reactions also require biotin. The citric acid cycle (Krebs) also uses biotin. There are several reactions in carbohydrate metabolism that require biotin. Protein metabolism uses biotin enzymes to break down amino acids, synthesize some amino acids and convert leucine in the liver. Creation of fatty acids also requires biotin reactions (McDowell, 2000, p. 451).

Requirements for dietary biotin depend on several factors, including the body's ability to make and absorb biotin in the digestive tract. Diet nutrients can increase need for biotin. Rancid fats increase the need but other antioxidant vitamins like vitamin E decrease the loss of biotin (McDowell, 2000, p. 453).

Food sources include organ meats, yeast, molasses, eggs, alfalfa, oilseeds and peanuts. Grains are low in biotin and other sources can be variable in their content. Not all biotin in food is available for absorption, however. Heat, milling, poor storage and extraction methods can destroy the vitamin (McDowell, 2000, pp. 454-455).

Deficiency

Deficiency leads to skin conditions, although the reproductive, nervous systems and adrenal and thyroid glands are all affected (McDowell, 2000, p. 455).

Ruminants

Deficiency leads to paralysis starting in the hind legs in ruminants. There is a link between potassium and biotin as well. Hoof health and condition relate to biotin. Although ruminants make and absorb biotin, the same problems associated with proper function of the digestive system and stresses to the animals that cause issues with the other B vitamins can lead to deficiency and need for supplementation of biotin (McDowell, 2000, p. 456).

In ruminants, newborns which are raised on bottle may need supplementation, but a real milk replacer (see recipe Chapter 11) can be used instead. Powdered replacers should be avoided at all cost. If the young have been raised on powdered replacers and need supplementation, 0.1 mg/kg or 100 micrograms subQ can be used for deficiency. Bacteria in the rumen that make biotin are very pH-sensitive, high grain diets will cause deficiency and increase acidity. D-biotin is the natural active form but degradation can occur if the vitamin is stored long periods or incorrectly (McDowell, 2000, p. 473).

Pigs

Like carnivores, pigs fed raw egg white may become deficient, and problems in digestion and absorption can interfere with biotin in the body. Deficiency symptoms include hair loss, slowed growth, poor skin, hoof problems, paralysis starting in the hind legs and lameness. Poor skin health and fissures in hooves can lead to secondary infections (McDowell, 2000, p. 457).

Poultry

Turkeys require more biotin than chickens, and symptoms in both species vary widely depending on when the deficiency starts. Poor growth and skin, leg and beak conditions and deformities with loose feathers are all symptoms associated with deficiency. Birds may have twisted legs and be unable to walk, particularly broilers. Pastured and naturally-raised birds, including broilers, circumvent the problems associated with deficiencies. Diets low in biotin mean poor and weak bone growth as well. Reproduction is affected and hatchability of eggs decreases with deficiency (McDowell, 2000, pp. 463-464).

Horses

The importance of biotin for hooves in horses is becoming very well-known, although if a horse is suffering from deficiency, it is worth investigating what may be causing this in addition to providing the needed biotin (McDowell, 2000, p. 464). In horses, supplementation to improve hoof conditions can be five mg (for ponies) up to thirty mg for draft horses, continued for several months. Biotin alone is not enough to grow healthy hooves, a proper mineral supplement and diet need to be provided as well, in addition to regular exercise. Since it takes a year for a hoof to completely regrow, care will be long-term, and proper management should continue for the life of the animal (McDowell, 2000, pp. 472).

Carnivores

Raw egg diets can lead to biotin deficiency, poor skin and hair condition. In cats, diarrhea, weight loss and anorexia also occur (McDowell, 2000, p. 466).

Biotin deficiency can be caused by any factor that interferes with synthesis or absorption in the digestive tract. This includes antibiotics and other drugs interfering with digestion, confinement systems (especially for those animals that practice coprophagy) or any management system that leads to stress and therefore digestive insufficiency. Mycotoxins, rancid foods and the genetic manipulation of breeds to increase growth also interferes with biotin absorption and synthesis. Frequent feeding of raw eggs to carnivores will cause deficiency (McDowell, 2000, p. 472).

Toxicity

Biotin is relatively non-toxic although very, very large doses can interfere with estrus cycles (McDowell, 2000, p. 474).

Folic Acid (Folacin, Folate)

The name for this vitamin (a group of several related compounds) comes from the word foliage (from Latin folium) since folacin is so prevalent in leafy green plants. Folacin is an orange-yellow crystalline powder that is not soluble in alcohol, is slightly soluble in hot water, and very soluble as a salt. It is more stable in alkaline and neutral substances, not as stable in acids. Oxidation and heat can destroy the vitamin, as can light and UV radiation. Preparation of foods is important in preserving the folacin content (McDowell, 2000, p. 483). Folacin can be found in all body tissues, especially the liver. Vitamin B12 is important for folacin metabolism and deficiency of vitamin B12 (possibly preceded by cobalt deficiency) leads to folacin deficiency as well. Unlike other B vitamins, folacin by-products are eliminated primarily in the feces (bile contains high amounts of folacin). During supplementation, high levels of folacin can be eliminated unchanged in the urine (McDowell, 2000, pp. 486-487).

Deficiency can be common in pregnant and lactating women, particularly in impoverished areas where diets are poor. Vitamin B12 and folate relate to cancer risk and high homocysteine levels. Homocysteine levels are better indicators for heart problems than cholesterol levels, which really bear little relation to heart attack risk. Ruminants synthesize their folate and vitamin B12; however, the same conditions that interfere with other B vitamin synthesis and absorption also occur with folacin synthesis. Pigs and poultry may need supplementation if raised unnaturally or with problems in digestion (McDowell, 2000, pp. 479-480).

Chemically, folacin is a structure called pteroylglutamic acid that also contains p-aminobenzoic acid (PABA). PABA is not required in the diet separately as long as folacin is in adequate amounts. The structure of folacin can contain many glutamic acid molecules, the synthetic form contains only one. There may be as many as 100 compounds of folacin in animals (McDowell, 2000, pp. 481-482).

Since cancer cells require high amounts of folacin for growth, substances that inhibit folacin are used as anti-cancer agents. Sulfonamides also inhibit PABA and therefore bacterial growth, acting as antibacterial agents (McDowell, 2000, pp. 483-484). Of course, this is somewhat backward thinking. Rather than try to treat these problems by inhibiting substances crucial to the body, how much better to prevent the problems with proper diet and treat using herbs and homeopathy that restore balance.

The body needs zinc to absorb folacin in the intestine and for conjugase (enzymes that break down folacin) activity. Conjugase activity occurs in the small intestine, liver and kidneys. The jejunum and duodenum of the small intestine are where most absorption of folacin occurs, by both active sodium transport and passive diffusion across membranes. Folacin is transported into cells in a monoglutamate form but restored to polyglutamate forms in cells. Folacin absorption can be very pH-dependent, folacin in citrus is less available because of the low pH (McDowell, 2000, pp. 485-486).

Folacin is stored in the liver in amounts that can last the body for almost half a year. Newborns quickly use up their stores and pregnant women need increased amounts. Folacin and vitamin B12 are usually supplemented in human pregnancies. Excretion occurs in feces and, in many animals, can exceed the dietary intake due to body synthesis of the vitamin.

Like other B vitamins, folacin is used as a coenzyme and in creation of fats, proteins, hormones, neurotransmitters, and in nucleic acid and amino acid synthesis.

Deficiency affects production of red blood cells and white blood cells and leads to anemia. Folacin deficiency causes incomplete transformation of the vitamin to active forms (like glutamate) and excretion of folacin in urine. Vitamin B12 is also needed for folacin metabolism and the ability of the body to move folacin derivatives across cell membranes. Methionine cannot be formed from homocysteine or choline from ethanolamine without vitamin B12.

The proper functioning of the immune system requires folacin, and antibody response is decreased in deficiency. Also, in deficiency of folacin, the pancreas does not function properly.

(McDowell, 2000, pp. 486-487, 489-490).

Ruminants should make enough folacin to meet dietary needs but again, this depends on proper digestion, lack of interference from other substances and other factors. Other animal species are at risk for deficiency. Diet can increase need for folacin or decrease synthesis, and high-protein diets in poultry have caused deficiency. Age is one factor that determines how much folacin is needed, older animals do not replace DNA as quickly, so need less folacin. Growth and pregnancy/lactation also influence how much folacin is needed. Newborn humans have better-functioning digestion when fed human breast milk rather than other milks and this can affect how much folacin is available to the infant. Medications can affect digestion and sulfa drugs cause deficiency, as do mycotoxins. Folacin utilization requires other nutrients; vitamins B12, C, choline and the mineral iron are all needed for proper use of folacin (McDowell, 2000, pp. 490-491).

Folacin is found in many fruits and vegetables, nuts and organ meats; however, the forms it is in varies. Eggs, milk and grains are not good sources. For herbivores, alfalfa, brewers grains, yeasts and hay that is put up and stored properly can provide decent levels of folacin. The body needs iron and vitamin C in order to use folacin. Light, heat and processing destroy the vitamin. Cooked vegetables should include the water they were cooked in to retain as much folacin as possible (McDowell, 2000, pp. 494-495). Pasture is a great source of folacin and, under most normal conditions, animals either make enough or take in enough through diet to meet their needs. In cases where medications like sulfa drugs, moldy feeds (corn and barley tend to always have some level of mold) or other problems either increase need or decrease absorption of the vitamin, supplementation is necessary (McDowell, 2000, p. 513).

Deficiency

Deficiency causes anemia and low white blood cell counts, in addition to problems with areas of the body where cellular growth and regeneration are rapid. These areas of rapid growth include skin, epithelial linings of digestive system, bone marrow, etc. Pigs and chicks require dietary folacin, as do primates. In carnivores, like dogs or rats, the body's synthesis of folacin usually meets needs, but there can be reasons this synthesis may not be enough and supplementation is required.

Ruminants

Healthy adults make enough folacin to meet their requirements but young animals without fully-functioning rumens are at risk for deficiency. Deficiency leads to low white blood cell counts, diarrhea, pneumonia and death.

Pigs

As with many other vitamins, feeding sulfa drugs and other medications or mycotoxins in feeds causes folacin deficiency (in pigs and other animals). The continuous feeding of antibiotics and coccidiostats will cause disruption in digestive function, decrease availability of key vitamins and minerals and eventually cause chronic disease and death.

Deficiency causes problems with immunity, reproduction (and number of offspring), anemia, bone marrow and other blood cell problems.

Poultry

Poor growth, leg deformities, lack of feathering and anemia associated with problems in the bone marrow are all symptoms of folacin deficiency. This can be quite common in birds not allowed out to pasture or fed diets low in folacin. Feather pigmentation also requires adequate folacin, iron and lysine. Egg weight, hatchability, weight of young birds at hatching and beak formation is affected by folacin, deficiency decreases all of the above. Choline and folacin interact, adequate choline means that birds need less folacin to prevent deficiency. Higher protein diets appear to increase need for folacin. On a cellular level, low folacin leads to lack of cell division and problems with the amino acids forming proteins.

Horses

Although horses synthesize their folacin in the digestive system, this may not be enough to meet their needs if they are kept off pasture and fresh greens. Lower performance is associated with deficiency.

Carnivores

Deficiency leads to anemia and low white blood cell counts, slower blood clotting time and inability to use iron. These all lead to loss of weight, lowered immunity and problems with chronic diseases. During chronic diseases and stress, folacin and all B vitamins are used up and deficiencies lead to continuation of the chronic condition.

(McDowell, 2000, pp. 495-504).

Toxicity

Toxicity is rare and is associated with injections of many times the normal dose. Pernicious anemia (vitamin B12 deficiency) from toxicity of folacin can go undetected until neurological damage is already started, so it is important to rule out vitamin B12 deficiency or to supplement the B vitamins in a complex form rather than as a single vitamin (McDowell, 2000, pp. 515-516).

Vitamin B12 (Cobalamin)

Vitamin B12 is completely linked to cobalt, the central molecule in the vitamin, and is a group of compounds and coenzymes. Deficiency produces pernicious anemia in humans. Pernicious anemia is now associated with vegans and some less careful vegetarians, although it can occur in other people for other reasons, like digestive dysfunction. Conditions that are a result of pernicious anemia include poor growth in herbivores and a weight loss syndrome in ruminants that has been termed "chronic wasting disease," "salt sickness" or "coast disease" (McDowell, 2000, pp. 523, 527). For information on cobalt, see Chapter 10 on Trace Minerals.

The cobalamin molecule can have one of several molecules attached that change the chemical formula; however, they all have cobalt as the base. Some forms found in manure may substitute in other minerals but these are considered inactive or even antagonistic to vitamin B12 (McDowell, 2000, p. 526). Cyanocobalamin is the form usually used in supplements because it is available for digestion and fairly stable. Vitamin B12 is soluble in water and alcohol, light and oxidation destroy it (McDowell, 2000, p. 528).

Absorption of the vitamin during digestion is a complex process. It starts with the saliva, which secretes a nonintrinsic factor protein that is bound to the vitamin B12 molecule after stomach acid and pepsin release the vitamin from the food protein that binds it. This B12 molecule bound to nonintrinsic protein cobalophin (from the saliva), passes into the small intestine, where a pancreatic secretion called trypsin acts to partially degrade the cobalophin and leave the vitamin bound only to an intrinsic factor, a glycoprotein synthesized in the gastric mucosal cells. All of this allows vitamin B12 to travel to the ileum where absorption occurs. This involved process requires proper saliva factors, gastric juices, pepsin and pancreatic function to release the enzyme needed to break the molecule apart. Any insufficiency in any part of the process can lead to deficiency (McDowell, 2000, p. 529).

Unlike the other water-soluble B vitamins, cobalamin is stored in body tissues, principally the liver but also the kidneys, brain, spleen and heart. Storage can be very high and help deal with times of deficiency. Excretion is by urine or feces. Vitamin B12 functions in the metabolism of nucleic acids, carbohydrates, fats and protein formation from amino acids. It is needed in maintaining the nervous system and creation of red blood cells. Vitamin B12 is part of the synthesis of nucleic acids. Thiamine is also part of nucleic acids that are needed for cellular function and formation. Deficiency in any of these B vitamins leads to problems in protein synthesis and cell division. Lack of B12 means that folate is not able to be properly metabolized, leading to hemolytic anemia and improper DNA synthesis, that has at its base either (or both) folacin and vitamin B12 deficiency (McDowell, 2000, pp. 532-533). Vitamin B12 is also important as an enzyme in the reaction of homocysteine to methionine, and deficiency leads to lack of protein synthesis from amino acids methionine, phenylalanine, serine and glucose. This lack of protein synthesis accounts for the weight loss or lack of gain seen in vitamin B12-deficient animals (McDowell, 2000, p. 534).

An important note about ruminants relates to the interaction of vitamin B12 and propionic acid, a by-product of carbohydrate metabolism/fermentation in ruminant digestion. Propionic acid in high levels leads to vitamin B12 deficiency and this is worsened when cobalt is deficient in the diet. Propionate accumulates in the blood and appetite is decreased. The appetite can be restored with injections of vitamin B12 or cobalt supplementation; however, the cobalt supplementation takes longer to reverse this anorexia. Cobalt deficiency cannot be adequately addressed with injections of B12 or injections of cobalt. Adequate levels of cobalt must be supplied in the diet. Because of these interactions, ruminants also have a higher requirement for vitamin B12 than other animals, as much as ten times the amounts needed in other species. Ruminants do not synthesize and absorb vitamin B12 as efficiently as they might and this also increases the need for vitamin B12 and cobalt (McDowell, 2000, pp. 535, 537, 538). Gestation, lactation and other physical stresses also increase needs. See below under Minerals for more information on cobalt needs.

Vitamin B12 is not needed in large amounts but is absolutely critical. The metabolism of vitamin C, folacin, methionine and choline are all interrelated. Animals with diets high in protein or those under stress require more vitamin B12. Vitamin B12 levels are related to pantothenic acid levels and to methionine levels, higher levels of one reduce the need for the others.

Many animals synthesize B12 in the intestine and can meet their requirements this way. The list of reasons this may be disrupted and lead to deficiency are the same as for the other B vitamins: stress, medications, diet, mycotoxins, inhibition of coprophagy or anything that interrupts proper digestion and absorption. Lack of cobalt in the diet is a huge factor in limiting vitamin B12 in the body, especially in those species that would otherwise have adequate amounts from intestinal synthesis. As was noted above, different feeds can cause problems in ruminants, higher carbohydrate diets or concentrated feeds limit vitamin B12 synthesis. Low-quality forage can be adequate for ruminants if they have proper amounts of copper and cobalt. Cobalt needs are possibly even higher than the National Research Council's requirements, the standard in established base nutrient requirements for animals (McDowell, 2000, pp. 535-537).

As many vegetarians and vegans have discovered, plants are not a good source of vitamin B12, in fact, they are terrible sources. A few exceptions include some seaweeds, although the seaweed does not make vitamin B12 but instead absorbs it from bacteria that become concentrated in the seaweed. Yeast and fungus do not, apparently, make vitamin B12, but digestive bacteria do. Ruminants should be able to meet basic needs with this synthesis, but other factors often interfere and lack of cobalt can definitely be a limiting factor. For carnivores, meats, milk (but not so much whey), eggs and organ meats (especially from ruminants) are very good sources.

For ruminants, cobalt is the critical limiting factor and deficiency can occur for various reasons. Forage may be low in cobalt if soils are low but also if soil pH is high. Alkaline soils do not allow plants to uptake cobalt well and rain can leach cobalt from soils. Because legumes need cobalt for their nitrogen-fixing bacteria in root nodes, they tend to be higher in cobalt than grasses (McDowell, 2000, pp. 538-540).

In people, vitamin B12 deficiency is pernicious anemia but in animals, it relates more to problems with weight loss (or lack of gain) than it does to signs of anemia.

Ruminants

Deficiency of vitamin B12 relates to either young animals before the rumen is fully functional, or to older animals with problems in digestion, absorption or factors interfering in either. Symptoms range from anorexia, weakness, nerve problems (demyelination of nerves) and low hemoglobin to anemia and low production of red blood cells in bone marrow. Sheep tend to be more susceptible than cattle or goats, but all ruminants can have problems associated with low cobalt (which determines the amount of vitamin B12 produced). Cobalt deficiency in soils is common in the U.S. and in tropical countries. Survival of lambs and resistance to parasites is also related to cobalt. Photosensitivity can occur in animals with deficiency, and deficient ewes may have fewer lambs and more abortions with weaker lambs at birth. Fatty or white liver disease is associated with cobalt deficiency, although the first sign is anorexia. It can be difficult to determine what is happening, a sheep may appear to be snuffling through hay but on closer inspection, the animal is only eating choice pieces and never a full mouthful. The cud is rarely chewed and extreme weight loss sets in. These animals tend to move away from the main flock or herd. This lack of herd interaction may be due to feeling poor, weakness or inability to compete with other animals. Malnutrition from cobalt deficiency looks just like malnutrition from lack of food, but the animals that are cobalt deficient may have plenty of rich feeds available and still lose weight, to the point of death. In fact, animals fed adequate cobalt make better use of poorer quality feeds. This chronic wasting and poor weight gain in cobalt-deficient animals is a major cause of production loss worldwide.

White or fatty liver disease is preventable with cobalt, as is *Phalaris* staggers, a disease caused by toxins in *Phalaris* genus grasses (like the reed canary grasses). *Phalaris* staggers may cause sudden death, but more often a staggering and trembling occur before seizures and death. Adequate cobalt may prevent this poisoning as well.

(McDowell, 2000, pp. 540-543)

Pigs

Pigs have similar deficiency symptoms to ruminants, with the addition of possible vomiting and diarrhea, poor skin and hair and loss of voice. Litter size and survivability of piglets is reduced, there may be abortions and deformities in piglets. Estrus can be delayed and fewer embryos produced. Each litter from a deficient sow may be weaker than the last, with more problems. Some animals exhibit anemia, some do not. (McDowell, 2000, pp. 544-545)

Poultry

Symptoms are similar in birds to what is seen in deficient ruminants and pigs. Poor weight gain, anorexia, nervous system problems and problems with feathering occur. There can also be weakness and bone deformities, but this may not occur early in the deficiency. Fatty liver, heart and kidneys and gizzard erosion can occur with deficiency. Since all the B vitamins are closely linked in their uses in the body, along with methionine, symptoms are similar for any of those that are deficient and enough vitamin B12 can compensate for deficiency of the others.

Although hens will still produce eggs, the egg size may be reduced in vitamin B12 deficiency, and hatchability is reduced as well. There can be problems with the thyroid in affected chicks in addition to the above symptoms occurring.

(McDowell, 2000, pp. 545–546).

Horses

Unlike ruminants, horses appear able to not only create the needed vitamin B12 in the intestine and absorb it, but they can do so at cobalt levels much lower than those required for ruminant health. Cobalt and vitamin B12 deficiency have not been a problem in horses.

Carnivores

In dogs and cats, young born deficient will not grow well and adults have reproductive problems. Symptoms are similar to other animal deficiency symptoms. McDowell reports in "Vitamins in Animal and Human Nutrition" (2000) that vitamin B12 deficiency has been found in half of German shepherds with degenerative myelopathy. Supplementation with the vitamin did not resolve the condition (p. 547).

Rabbits

If allowed to practice coprophagy as they normally would, rabbits should not have vitamin B12 deficiency.

Unlike many other vitamins, vitamin B12 is fairly stable, even in mineral mixes, and is often supplemented as cyanocobalamin. One of the biggest problems is not stability of the vitamin, it is inability of the animals to absorb it. This may be because the animal already has sufficient stores (or synthesizes it) or because poor digestion inhibits absorption.

In cases where digestion is impaired, for whatever reason, injection of the vitamin becomes necessary. Injectable B12 is usually as part of a complex with other B vitamins, but it can be purchased as a single injectable. Rates of injection differ slightly by species. One hundred micrograms once a week for lambs or 500-3000 micrograms per cow per week can correct deficiency brought on by cobalt deficiency. This, however, does not meet the needs of cobalt in ruminants, in addition to that required to make vitamin B12. Making sure animals get enough cobalt makes injections unnecessary under ordinary conditions (McDowell, 2000, pp. 546-556).

Some people now use pellets or boluses of cobalt in various forms. This has some drawbacks, the pellet may pass through the digestive system rather than staying in the rumen/reticulum or it may become coated with calcium phosphate, making absorption of cobalt impossible. Some pellets are more prone to be ejected during cud chewing as well (Ammerman, Baker and Lewis, 1995, p. 121).

Cobalt deficiency affects different ruminants within a flock or herd differently, and weather, stress or other factors also play a role. Some years all animals do well, some years some animals do well and others are marginal. Some years many animals are affected and some may even die. This variability in deficiency makes it difficult to tell exactly what is happening. Supplying cobalt continuously is one good way to ensure animals always get what they need (McDowell, 2000, pp. 556-557).

A note about humans: vegetarians and vegans should take supplemental vitamin B12 since their diet may not provide enough (particularly for vegans).

Toxicity

Like the other B vitamins, excessive dietary vitamin B12 is non-toxic in very high amounts. Cobalt can be toxic in high levels in ruminants although the margin of safety is quite high. Toxicity produces shortness of breath, excess urination, salivation and increased red cell counts in feces. High levels of cobalt are usually due to mixing errors in mineral supplements rather than from natural sources (McDowell, 2000, p. 558). Other species' tolerances differ, but their need is not as critical as that of ruminants.

Vitamin C (Ascorbic Acid)

Most animals make their own vitamin C, with the exception of guinea pigs, primates (including humans), some birds and fish. This is why, if you feed commercially prepared pellets to your guinea pigs (with no fresh vegetables or fruits), you must get a formula specific to guinea pigs. The rabbit formulas do not contain vitamin C and guinea pigs must have this in their diet. Feeding a real foods diet to these animals will prevent deficiency and is a healthier option.

For many years, recommended amounts of vitamin C only included amounts needed to prevent scurvy, the disease associated with complete deficiency. It is now generally accepted (at least by natural health practitioners) that vitamin C is needed in much much higher amounts to ensure health. Vitamin C, like vitamin E, selenium and vitamin A, is an antioxidant that helps remove free radicals from body tissues. Anything in the body or environment that increases stress, exposure to toxins or problems with digestion that create metabolic toxic wastes increases the need for these vitamins and minerals (McDowell, 2000, p. 597).

Scurvy has been described as early as 1500 B.C. in humans. Native Americans drank infusions of spruce needle tea to prevent symptoms during winter although European peoples suffered from deficiency during winter when vegetables and fruits were not available. During sea voyages, deficiency was a huge concern and accounted for many deaths during wartime as well. Eventually, navies realized the link between citrus (lemons and limes) and scurvy prevention. By 1795, lemon or lime juice was part of a British sailor's ration, leading to the term "limey" for sailors. Even by the early 1900s, bottle fed babies were still at risk for scurvy because milk was boiled and this destroyed the vitamin C (McDowell, 2000, pp. 599-600). Unfortunately, scurvy can still be a problem, even in developed countries. In third world countries, drought, poverty and poor nutrition still contribute to scurvy. In developed countries, poverty, poor eating habits, processed foods and poor standards of care for children and elderly mean scurvy is still a problem.

Vitamin C has two primary forms: ascorbic acid and dehydroascorbic acid (oxidized from ascorbic acid). In commercial uses, erythorbic acid (an isomer of ascorbic acid) is used to keep foods from discoloring during processing and exposure to air. This property of ascorbic acid to oxidize to dehydroascorbic acid and back again accounts for vitamin C's antioxidant property, and also means the vitamin is the least stable of all vitamins and most easily destroyed during processing.

Vitamin C is very water-soluble and acidic and tends not to be available in dry foods or in cooked foods, especially if there is alkalinity. Environmental toxins increase needs for vitamin C and act as antagonists to it. Examples include heavy metals, tobacco smoke (cigarettes are heavily laced with toxins and heavy metals), diuretics, antidepressants and other drugs, industrial toxins and other pollutants (McDowell, 2000, pp. 600-601). Anyone choosing to smoke should consider the increased need for vitamin C and supplement their diet (or better yet, quit smoking). The role as antioxidant that vitamin C plays is not understood completely but it is essential for this function and several others discussed below.

The body digests vitamin C as though it were a carbohydrate although passive diffusion can also occur. Low levels of the vitamin are absorbed through an active sodium transport system, higher levels through diffusion. Even though vitamin C is not common in many foods, it is well-absorbed. Where the absorption takes place is dependent on the species to an extent. Rats absorb vitamin C in the ileum while guinea pigs (which do not manufacture their own) absorb it from the duodenum.

Although vitamin C is not stored appreciably in the body, the adrenal and pituitary glands do store some vitamin C and it is found throughout the body in tissues, muscles, organs and the brain. Low-level adrenal fatigue and adrenal imbalances (very common in America) benefit from vitamin C. Wounds accumulate vitamin C and stress reduces the body's vitamin C levels and increases the synthesis of the vitamin in those animals that make their own. In people, vitamin C degrades by half in about two weeks to a month, in guinea pigs it can be as short as three days. People not getting any vitamin C can develop scurvy in three months but low level deficiency may be quite common in elderly people, chronically ill, stressed individuals or those exposed to toxins continuously. Most vitamin C is excreted in the urine although lagomorphs (rabbit family) and rodents may lose vitamin C through carbon dioxide. Excretion from the kidneys in urine depends on glomerular filtration rate. This is why doctors often say people with kidney failure cannot take vitamin C. While supplementation with high amounts may be problematic, these individuals do need vitamin C and should watch their diet carefully to include plenty of vitamin C-containing fruits and vegetables (McDowell, 2000, p. 603).

Collagen

Collagen synthesis and repair in the body requires vitamin C. Collagen fibers are proteins found in skin, teeth, bones, connective tissues (ligaments and tendons) and between cells. Their formation involves the amino acids lysine and proline. Vitamin C also helps cells differentiate during formation by altering gene expression. The vitamin C may act to keep iron in a ferrous state so that hydroxyproline can be formed from proline and used to synthesize collagen. Without these reactions, collagen stays in a precursor state that is not fibrous or supportive and the individual develops scurvy. Collagen is also used in the body at wound sites, and vitamin C supports proper healing and capillary strength anywhere in the body. Problems with wound healing, gum and bone disease are related to low vitamin C (McDowell, 2000, pp. 604-605).

Antioxidant

The antioxidant effects of vitamin C are perhaps the most well-known functions of any vitamin but are poorly understood. Examples of free radicals in the body include hydrogen peroxide, hydroxy, peroxy, alkoxy and superoxide and are created by normal metabolism, radiation or actions of immune system cells when they destroy pathogens. Oxidative damage can be a result of normal biological function unless antioxidants can remove those damaging free radicals before they injure healthy cells. Substances that act as antioxidants and protect cells from damage include vitamin C, vitamin E, vitamin A and minerals iron, selenium, zinc, copper and manganese. Of these, vitamin C is the most important, mostly due to the fact that it so readily oxidizes and then reduces, taking other free radicals with it. Any time there is illness, injury or stress, the body's need for these nutrients increases substantially. The antioxidant nutrients also help support immune function, deficiency in any of them increases risk for infections and lowered immunity. Lowered risk for cardiovascular disease and cataracts is also related to antioxidant function. Vitamin C in particular stimulates immune cells that protect against viruses and cancer. Vitamin C deficiency is related to cancers of the digestive system. Vitamin C also reduces glucocorticoids, which are cortisols produced by the adrenal glands that reduce inflammation but also lower immunity. Vitamin E can be revitalized by vitamin C and total cholesterol can be lowered if there is enough vitamin C. The oxidation of LDL cholesterol is inhibited by vitamin C (McDowell, 2000, pp. 605-607).

Vitamin C also helps keep metal ions in a useable form in the body. This means these metal ions are more available for absorption and transport into cells, like copper and iron. Vitamin C also protects cells from toxic metals, like lead, and helps remove lead from the body.

The relationship of vitamin C to amino acids includes the need for ascorbic acid in amino acid carnitine formation from lysine and methionine. Carnitine can prevent accumulations of triglycerides in blood, improve fat metabolism and functions as part of the citric acid cycle (Krebs).

The B complex of vitamins and vitamin C function together in many metabolic reactions. Folacin and vitamin B12 synthesis and use require vitamin C and deficiencies of the B vitamins cause problems with urinary excretion of vitamin C.

Vitamin C may also inhibit aldose reductase, the enzyme that reduces glucose to sorbitol. This is important in diabetics, who may benefit from supplemental vitamin C to reduce sorbitol levels.

The toxicity of nitrosamines as carcinogens is now well-known and vitamin C plays a role in reducing formation of nitrosamines. The body also needs vitamin C to detoxify from metabolic molecules and wastes, from drugs and to support the body's immune system in diseases. Fertility, particularly in the male, depends on vitamin C as well (McDowell, 2000, pp. 607-609).

As was mentioned above, most animals can synthesize vitamin C, with the exceptions of humans and other primates, guinea pigs, some birds, fruit-eating bats, fish, some reptiles and insects. Even in species that can make their own vitamin C, however, some individuals may not make or absorb enough to meet their needs, and in pigs and rats, this may be genetic. This ability to synthesize vitamin C depends on an enzyme, L-gulonolactone, that is missing in the above species. Although nutrient requirements for vitamin C in species that make their own is not set, nevertheless, there may still be a need for vitamin C in some animals, and supplementation or access to foods high in vitamin C is strongly recommended. In humans, there are several factors that increase need for vitamin C. Elderly people have an increased need, as do pregnant or lactating women. People with poor diets, poor digestion, those with hyperthyroidism, those who abuse alcohol or drugs and those using oral contraceptives also have a greater need for vitamin C.

Requirements for animals vary widely and those for humans even more so. One school of thought for humans is that supplementation should prevent scurvy. To prevent scurvy only requires about 10 mg vitamin C daily. To support the other metabolic and immune functions of vitamin C requires much more and to aid in detoxification, disease treatment or prevention or other physiological support requires high levels on a daily basis. Recommendations may vary from the RDA (sixty mg for an adult) to ten grams daily for specific purposes and individuals. These high levels have shown positive effects, including preventing colds and disease (including cancer), increasing healing and immunity, treatment of viruses, high cholesterol and atherosclerosis, treatment of disorders like schizophrenia and prevention of megaloblastic anemia in bottle-fed babies (which can be caused by vitamin B12 deficiency) (McDowell, 2000, pp. 610-611).

Most people are familiar with the higher vitamin C content in citrus fruits but vitamin C is also found in other fruits and vegetables (including potatoes, cabbage cauliflower, and even some animal products, like fish). One of the highest sources of vitamin C is rose hips (*Rosa* spp.), however, species of rose, time of year and storage affect available vitamin C. Rose hips and some other berries and tea leaves (fresh) accumulate vitamin C. Vitamin C content can vary by part of the fruit or vegetable, how it was processed, how many enzymes are in it and temperature. Oxidation destroys vitamin C, so fresh sources are best. Other causes of destruction include copper and iron content, long cooking times at high temperatures and oxygen. Heating quickly can be beneficial though, because it destroys enzymes that otherwise break down the vitamin C. Dehydration and freezing can be good options for preserving foods with vitamin C content intact (McDowell, 2000, pp. 612-613).

Deficiency

Generally, those animal species that make their own vitamin C are not at risk for deficiency. In fact, dietary vitamin C has little effect on their body levels. Problems in the digestive system, stress, poor nutrition and low vitamin A (or beta carotene) intake can cause deficiency, however (McDowell, 2000, p. 613).

Ruminants

All ruminants make their own vitamin C. In instances where something does cause deficiency, symptoms in young animals include problems with the mucus membranes of the mouth and dermatitis. Weight and hair loss, thickened skin and poor milk production and quality also occur with deficiency. If there is a problem in ruminants that inhibits synthesis, dietary intake does not work well because the vitamin is destroyed in the digestive system. Giving large amounts of oral vitamin C can be somewhat beneficial but injection becomes the only reasonable form of supplementation. Another reason to support proper digestion and absorption (McDowell, 2000, p. 614)!

Very young animals (particularly ruminants, like calves) less than three weeks old may have lower vitamin C levels in the body than adults, especially if they did not get enough colostrum. Animals with infections or problems in the digestive system have lower levels of vitamin C. Tapeworm infection in cattle reduces the amount of vitamin C in the body (McDowell, 2000, pp. 614-615).

Pigs

Deficiency in pigs can occur during times of stress (including environmental stresses), disease or very young animals. Low energy in the diet affects amount of vitamin C in tissues. Milk is high in vitamin C so nursing piglets have enough, but at weaning may be at risk for deficiency. This is particularly important if piglets are moved or handled at weaning, and therefore stressed. Under normal conditions, pigs are able to synthesize enough vitamin C to meet their needs. Symptoms of deficiency include weakness and pain in bones, and skin, muscle, fat and organ hemorrhages. Legs may become deformed due to weak tendons and ligaments and boars produce less sperm. Deficiency may also be genetic, which can arise naturally but is very often a result of breeding for confinement conditions. Confinement-raised animals may also have higher requirements for vitamin C (McDowell, 2000, pp. 615-616, 629).

Poultry

Most birds, like ruminants and pigs, make their vitamin C but stress (including heat stress), vaccination, diseases, temperature swings and young birds' inability to synthesize as much vitamin C all contribute to possible deficiency. Low levels of the vitamin affect egg production, fertility and susceptibility to disease. Vitamin C is important in calcium metabolism and use of vitamin D in the body. As with any animal, dietary sources are the easiest way to supplement vitamin C and provide good levels without excess. Chickens readily eat all types of fruits and vegetables and do well in a pasture or free-range system (McDowell, 2000, pp. 617-619).

Horses

Horses also synthesize vitamin C, but again, stress and disease increase need for the vitamin and can lead to deficiency. Deficiency affects sperm, fertility in the mare and immunity (McDowell, 2000, p. 620).

Rabbits

Rabbits synthesize enough vitamin C to meet their needs and should not require a dietary source (unlike guinea pigs) (McDowell, 2000, p. 624).

Carnivores

Carnivores do synthesize their vitamin C but at much lower rates than herbivores. Supplemental vitamin C may be beneficial in joint problems, like arthritis and hypertrophic osteodystrophy, and in treatment of disease. In cats, damage to the central nervous system can result in large loss of vitamin C (McDowell, 2000, p. 620).

Herbivores

Under normal conditions, herbivores with access to pasture should not need supplemental vitamin C. Any type of stress or improper diet, however, runs the risk of causing deficiency. Animals kept in confinement will be deficient in all the antioxidant vitamins and minerals and become susceptible to all types of diseases.

Young are born with high stores of the vitamin but will need immediate sources, like colostrum and milk. Lack of colostrum at birth contributes to deficiency, disease and death. Milk replacers (unless based on real milk) will most likely not provide the necessary vitamin C. Weaning of young early leads to deficiency, at least partly from stress (McDowell, 2000, pp. 627-628).

Vitamin C is available as a supplement in several forms, the crystalline form is an easy-to-use form that is fairly stable. Some products coat the crystals (ethylcellulose) for even greater stability. Several other forms have improved stability and can be used when available. Natural sources are better because they also contain other bioflavonoids; however, stability of vitamin C is a big issue and care must be taken when preparing, storing and using plant sources (McDowell, 2000, p. 630).

Toxicity

Vitamin C is generally considered fairly safe at higher levels but continuous high-level supplementation may lead to interference with digestion of minerals, irritation of the digestive tract, acidosis and sensitivity. In humans, concerns about higher levels of supplementation include formation of uric acid crystals in the kidneys and increased absorption of iron (leading to iron toxicity). These effects are not usually seen because vitamin C is so rapidly eliminated from the kidneys. People with kidney failure may be advised by doctors to avoid supplementation, however, food sources of vitamin C can be beneficial.

Animals appear to tolerate high doses of vitamin C very well with no adverse effects (McDowell, 2000, pp. 633-634).

CHAPTER 9

THE MINERALS

The information on minerals pertains, in particular, to herbivores, not because carnivores do not require minerals but because carnivores generally meet their mineral needs from diet alone. Herbivores are at the mercy of the pasture where they are kept or the hay they are fed and their mineral needs often depend on some type of additional supplement.

A note about plants and minerals: not all plants in a pasture contain the same amounts of minerals. Plant species' differences and stage of growth, whether the mineral is available in soil, weather and seasonal differences all affect how minerals are taken in by plants, and therefore available to animals grazing. If a particular mineral is not in the soil where the plants are grown, the plants cannot be a source of this mineral! Biodynamic theories consider transmutation of elements as possibilities but I believe this may prove to be somewhat erroneous. Our understanding of the complexities of life, metabolism and the interaction between soils and plants, and plants and animals is not beyond elementary at best. It is entirely possible we do not yet understand the complexities of mineral availability. Plants can pull up minerals from substrate. Changes in weather, pH of soils and diversity within plant communities determine what minerals are available to plants. Even mycorrhizal connections and the health of bacterial populations in soils play a role in how plant roots can use minerals.

There are tests available for forage and a general analysis of mineral and content of plant material. These tests allow farmers and manufacturers to gauge how much and of what type of supplemental minerals will be needed on a particular farm during a certain season. But the tests have limits and cannot gauge with any accuracy the true health of the microbiome of the soil. The health of plant communities and animals grazing them can give a better idea of whether your soils are healthy. Just as you can look at an animal and see health in bright eyes, shiny coats and clear respiratory tracts, you can look at plant communities and grazers on those plants to see signs of soil health. I no longer advise applications of minerals to soils as this changes the soil and plant communities from what is natural and native to an area. Now I advise that people look toward biodynamic or homeopathic preparations that help restore proper mycorrhizal, bacterial, insect and plant communities. Addition of composted manure from the farm where animals are kept in a natural manner and fed proper nutrition can provide the trace minerals back to soils in a way that is useable by plants and therefore available to herbivores.

The importance of maintaining pastures with as much diversity as possible, including some shrubby species where appropriate, cannot be overemphasized. This diversity and availability of browse not only provides the widest variety of minerals and nutrients possible, it can also help with parasite control (discussed in volume two). Do not underestimate the benefits of native plant communities for herbivores. These communities are adapted to local soil and weather conditions and have often adapted to grazing and browsing of native herbivores. Many of the native herbivores may have been extirpated or, conversely, occur in higher-than-normal numbers now due to human mismanagement of resources. Rotational grazing allows a farmer to maximize pasture or browse without doing permanent damage to the plant communities because animals are moved off before overgrazing and destruction can occur. Many wild herbivores maintain large home ranges and roam for many miles in the course of their lives. Replicating this benefits the animals and the plant communities.

Because most farming situations involve fencing and a limited area for animals, these animals cannot follow their instincts and search for the minerals they need by roaming for miles to areas of different soil. The farmer must provide missing minerals and hopefully, improve soils over time. Grazing animals have the potential to increase soil fertility and slowly improve mineral content through their manure.

Applications of composted manure to hay ground help restore soils by not only adding minerals, but by providing proper conditions for the many beneficial bacteria, fungus and insects that are crucial to soil health. The soil is an ecosystem that maintains whole communities of organisms at the microscopic level as well as the level of insects. In turn, these organisms and insects work the soil during their life cycle, moving and freeing up nutrients for plants to use, supporting plant health with symbiotic relationships and thereby supporting the herbivores that graze the plants. Each farm is unique and each part of the country or ecosystem has its own populations of supporting players in the world of soil health. Working with the land is an easier option than trying to subdue organisms that become harmful when they flourish out of balance in a damaged ecosystem.

For many of us, the process of restoring the soil and plant communities to a healthy state that can support grazing animals is ongoing and may have started from a negative. While the restoration takes place, animals still need proper nutrition and this may not yet be provided by the pastures. Smaller farms with limited land options may not be able to ever support herbivores without some supplemental minerals. In these cases, providing minerals to the animals is the only reasonable way to ensure health. Following is information on each of the main minerals needed but it is not an all-inclusive list. There are many trace minerals needed for health not included in this book. Food sources are the best source of these, including organic kelp, pasture plants, browse species, fruits and vegetables.

Macro Minerals (Calcium, Magnesium, Phosphorus, Chlorine, Potassium, Sodium, Sulfur)

These are the building blocks of the body, minerals needed in high amounts for daily physiological processes. Imbalances in one of these minerals leads to imbalances in all the related minerals. An example to illustrate the importance of considering minerals as complex and interconnected parts of a living organism:

A ewe presents with a prolapse a few weeks from lambing. Since calcium is the mineral that allows muscles to contract, low calcium can lead to weak muscles that are unable to hold in tissues when there is pressure (as from lambs). Prolapse (either uterine or rectal) can result. This is the first clue that this ewe does not have access to enough calcium (or the related minerals or vitamins needed to absorb calcium) and right at this moment, the problem should be addressed. As time goes on, this ewe develops milk fever. Milk fever can occur in sheep in late pregnancy, in dairy cattle it usually occurs after lactation has started. The veterinarian is called and calcium injections are prescribed. Because calcium needs to be in a ratio with magnesium, larger amounts of calcium supplementation should also be accompanied by magnesium or else the ewe will next develop pregnancy toxemia. The toxemia leads to death of the lambs in utero and subsequent septicemia. Without immediate attention (and sometimes even with it), the ewe will die, as well. Her low mineral levels also make it difficult for uterine contractions to expel the dead fetuses.

Preventing the above is obviously easier than treating. It is important to remember that some animals have a genetic propensity to not assimilate nutrients as well. Some animals have poorly-functioning digestion and assimilation due to faulty management, poor quality feeds, stress, illness or other factors. Animals cannot take in what is not there. Without access to the proper minerals in soils, forage, hay or mineral supplements, animals will become deficient.

Herbs and homeopathy offer ways to help the body make use of what it has, and in the case of herbs, to provide bioavailable forms of some key nutrients along with the supportive vitamins and minerals. These therapies will be discussed in depth in volume two.

Calcium (Ca)

Calcium, magnesium and phosphorus, although in separate sections, really should not be divided out as individual minerals. Their uses in the body are interrelated and ratios between each of them are as important (if not more so) than actual amounts in the diet.

The amounts of calcium needed for health depend on stage of growth of the animal, pregnancy or lactation. Needs also depend a great deal on the availability of related minerals (like magnesium, phosphorus and boron, among others) and vitamins, like vitamins A and D. A daily average for most livestock is between one half and one percent calcium, laying birds require as much as three and a half percent calcium daily (NRC, 2005, p. 97).

Calcium is the most abundant mineral found in the body, with up to ninety-nine percent found in the bones and teeth. Thirty-five percent of the content of bones is calcium (Underwood, 1981; NRC, 1989, p. 10, Coleby, 2010, p. 17). The body uses calcium to maintain proper pH in the blood. If blood levels of calcium drop, parathyroid hormone converts vitamin D to calcitriol (another hormone). With calcitriol, calcium is absorbed from the small intestine and the bones and moved back into the blood (NRC, 2007, p. 114). The body excretes calcium in both feces and urine (NRC, 2007, p. 114).

Calcium needs to be considered in ratio to phosphorus and magnesium in the diet, imbalances in these ratios lead to urinary calculi and other health issues. Excesses of calcium can cause problems with absorption of several other minerals. Vitamin D and trace minerals, like boron, are needed along with calcium for proper absorption in the body (Coleby, 2006, p. 92; Abrams, 2000, p. 67-68, Pugh, 2002, p. 23). Calcium needs to be in a ratio with magnesium of 2:1 and calcium to phosphorus approximately 2:1 or 1:1 (depending on life stage) (Abrams, 2000, p. 68).

Mentioned previously under Vitamins, it is worth repeating that vitamin D, while it occurs in forage, is most commonly made by the skin in herbivores but needed as a dietary source in carnivores. Animals grazed in northern latitudes or housed in barns for part of the year may become deficient (Underwood, 1981, p. 32).

Calcium is often found as calcium carbonate in soils but can be added to supplements as dolomitic limestone or any of several organic forms of calcium, including calcium citrate (NRC, 2005, p. 97). Availability for digestion depends on the coarseness of the compound, finer-ground calcium is better absorbed and organic forms are very well-absorbed. Calcium is available from forage, particularly alfalfa. The soluble salts are more bioavailable; however, oxalates can interfere with absorption of these. Adequate vitamin D can compensate for some of the oxalate problems (McDowell, 2000, p. 113).

Calcium is used in the skeletal, nervous and muscular systems, for proper heart function and blood coagulation. The body stores calcium in the bones and may draw on this reserve for other uses during times of dietary lack. In prolonged periods of deficiency, osteomalacia or osteoporosis will result (Abrams, 2000, p. 69). Calcium is also used in teeth and nails and gives vitality. It promotes healing of wounds, reduces acidity in blood and the digestive system, prevents faulty food assimilation and safeguards the health of developing embryos (Levy, 1991, p. 170).

Deficiencies

Like all other minerals, calcium cannot be divorced from the minerals that relate to it and the body's ability to use calcium. It is linked to deficiencies of the related minerals and vitamins, like magnesium, phosphorus, boron and vitamins A and D. Some plants can reduce availability of calcium, one percent oxalates in the diet reduces the calcium available by sixty percent. Phytates also interfere with absorption (NCR, 1989, p. 10).

Differing life stages can increase the need for calcium. Growing animals and lactating females need higher levels of calcium. Grains are low in calcium so young animals fed high amounts of grains can become deficient and show signs of rickets, leg deformities and urinary gravel (NRC, 2007, p. 116).

Prolonged deficiency or malabsorption can lead to arthritis, uneven bone growth, enlarged joints, osteoporosis, pregnancy hypocalcemia, muscle weakness (leading to prolapses), decreased milk production, cow hocks and poor teeth or teeth that fall out prematurely. Teeth that fail to erupt can also relate to calcium and related vitamin and mineral deficiencies (Pugh, 2002, p. 23; Coleby, 2006, p. 93; Abrams, 2000, p. 69). It is important to note here that excess of some minerals, like phosphorus, will cause problems with calcium absorption and present as calcium deficiency when, in fact, the problem is one of excess of antagonistic minerals. In humans, carbonated sodas that contain phosphates contribute to low calcium and magnesium. Most American diets contain enough calcium but they may not contain enough magnesium.

Mastitis is linked to calcium imbalance (high calcium/low magnesium) as well as to low selenium and copper. A general lack of well-being and susceptibility to colds and respiratory problems can also relate to imbalances in calcium as well as selenium, vitamins E and A, copper and manganese (the antioxidant vitamins and minerals).

(Coleby, 2006, p. 93)

Toxicities

Excess calcium throws off ratios of magnesium and phosphorus and interferes with other trace minerals. This can lead to urinary calculi, risk of which worsens in areas where cold winters keep animals from drinking enough water. Mastitis due to incorrect ratio with magnesium can occur and overall immunity is lowered (Coleby, 2006, p. 93-94).

It is difficult to determine calcium levels in the body by blood serum testing. As was mentioned above, the body will take calcium from the bones to balance blood levels in order to keep blood pH correct for health and life.

In most animals, once calcium needs are met, the body stops absorbing calcium from the diet but rabbits and horses continue to absorb calcium in excess of their needs. This absorption is independent of vitamin D levels in the body (NRC, 2005, p. 99). Low-level increases in calcium over time do not usually cause toxicity symptoms, although this can interfere with absorption of other minerals, including phosphorus and zinc (and imbalances with magnesium) (NRC, 2005, p. 100). If calcium is forced in high levels (such as with a bolus or drenches given to animals to prevent hypocalcemia of birth and lactation), toxicity can result. This toxicity results in acidosis, possibly tissue calcification and eventual death. The form of calcium given makes a difference; the propionate (chelated) form can cause toxicity, while the mineral form tends to be safer, possibly because it is less well-absorbed (NRC, 2005, pp. 99-100).

Pigs

When calcium is too high and out of balance with phosphorus, phosphorus absorption is low and bone abnormalities occur. Ratios of calcium to phosphorus close to 1:1 have the best chance of proper absorption of each mineral (NRC, 1998, p. 47). It is important to remember that healthy levels of minerals that provide nutrition for proper organ function and bone structure are not necessarily the amounts that give fastest growth rate. Pigs, in particular, are managed improperly, for fast weight gain for market rather than healthy, slower-growing animals that have access to pasture and proper mineralization (NRC, 1998, p. 48). Boars and somatotrophin (a growth hormone) treated pigs have higher mineral needs than castrated animals (barrows) or females (gilts). Pig breeds that tend to be more lean do not have a greater requirement for calcium or phosphorus (NRC, 1998, p. 48). Grain-based phosphorus (phytate) is not readily available for digestion in pigs. Feed and mineral manufacturers now add phytase, the enzyme needed to digest phytate. Phytase also increases the bioavailability of calcium for pigs and can make a drastic improvement in their ability to digest both calcium and phosphorus (NRC, 1998,

p. 49). Wheat may be a better grain for pigs since wheat also contains phytase (NRC, 1998, p. 48). Alfalfa is a good source of both calcium and phosphorus and, in pigs, meat and bone sources are also highly digestible. Remember that pigs are omnivorous, their digestive system makes use of both vegetable and animal food sources (NRC, 1998, p. 126).

High levels of either calcium or phosphorus can reduce weight gain and interfere with absorption of other minerals, including zinc, and increase need for vitamin K (NRC, 1998, p. 49).

Horses

As is the case in other animals, calcium must be in balance with phosphorus for health. Calcium deficiency leads to rickets in young animals and bone problems in older animals. Alfalfa hay may be too high in calcium for horses that do not also have a source of phosphorus in the correct ratio to the dietary calcium (NRC, 1989, pp. 10-11).

Carnivores

Phosphorus and calcium are considered together, as they are in all animals since the ratio is more important than actual amounts of either mineral. A 1:1 ratio is approximate in most mammals and in dogs, the proper amount and ratio of both minerals can compensate for low vitamin D. The size of the dog makes a difference in requirements of each mineral, larger dogs need more calcium and phosphorus for proper bone growth. Deficiencies lead to rickets in all dogs, larger breeds are more susceptible because of their greater need (NRC, 1985, p. 15). The need for raw bones as part of a healthy carnivore's diet is an important consideration. The ratio of calcium to phosphorus is important and studies have shown that improper ratios lead to hyperactive parathyroid gland and poor tooth health (including teeth falling out) (NRC, 1985, p. 16). Higher amounts of calcium and phosphorus than requirements can also cause problems. Large dogs like Great Danes have been shown to have bone abnormalities when fed diets too high in calcium, phosphorus and vitamin D (NRC, 1985, p. 16).

Plant sources

Alfalfa and other legumes are common sources of calcium but plant sources are actually not as well absorbed as mineral sources for calcium. Absorption of calcium from minerals can be up to 90% (NRC, 2005, p. 99). Other plants that contain calcium include lambs quarter (*Chenopodium album*), nettle leaf (*Urtica dioica*), mulberry leaves (*Morus* spp.), neem leaf (*Azadirachta indica*), sumac (*Rhus* spp.), German or Roman chamomile (*Matricaria recutita* or *Anthemis nobilis*), chicory (*Cichorium intybus*), cleavers (*Gallium* spp.), dandelion (*Taraxacum officinale*), horsetail (*Equisetum* spp.), common or English plantain (*Plantago major* and *P. lanceolata*), shepherd's purse (*Capsella bursa-pastoris*), willow (*Salix* spp.) and others. Plants generally also contain the trace minerals and vitamins needed for proper utilization of the calcium. If, however, soils are deficient or mineral imbalances occur in soils, plants may not provide the needed mineral levels (Levy, 1991, p. 170; Walters, 2013, pp. 192-193).

Magnesium (Mg)

Amounts of magnesium needed depend on life stage and ratio needed to balance calcium and phosphorus. Magnesium is usually obtained from grazing, but highly fertilized pastures and grains may be deficient. Type of soil makes a big difference in availability as well, sandy soils do not hold magnesium well.

Magnesium carbonate, oxide and sulfate are common forms used in livestock feed. The more finely ground the magnesium, the better it is absorbed by the animal (NRC, 2005, p. 224). The magnesium oxide form available in dolomitic limestone (that also contains calcium in proper ratio to magnesium) is a good source of magnesium, particularly for non-ruminants (NRC, 2005, p. 226). When a more bioavailable form is needed (in horses with chronic laminitis, for instance), magnesium citrate can be used. Other chelated forms include magnesium aspartate but aspartate is a neurotransmitter in the central nervous system. Whether or not dietary aspartate is able to cross the blood-brain barrier is questionable but until more is known about the effects of excess aspartate on brain function, it is better to use a different form of chelated magnesium.

Animals and people low in magnesium become nervous (hyper-excited), people become anxious and the blood pressure rises due to physiological processes. Some natural health practitioners estimate that much of the anxiety and hypertension in the U.S. today is due to magnesium deficiency rather than some underlying disease.

The small intestine is the site of magnesium absorption in almost all animals, with the exception of ruminants (see below). Unlike calcium and phosphorus that are removed from bone by the body to meet needs during dietary deficiency, magnesium is not easily moved from the bone matrix. Dietary sources are necessary and need to be fairly constant, excess is removed from the body by the kidneys. The kidneys can recycle magnesium back into the body when needed in response to the parathyroid releasing hormones. The parathyroid excretes hormones in response to hypocalcemia but not to hypomagnesemia and both minerals (calcium and magnesium) are then recycled from the kidneys back into the body (NRC, 2005, p. 225). Since the body only responds to low magnesium by increasing reabsorption of both calcium and magnesium, the body has no way to compensate for low magnesium.

Soil magnesium is available for plants and thence for animals, but the availability decreases in cold, wet weather and if potassium is high (NRC, 2005, p. 226).

Magnesium is needed for normal nervous system functioning (calms nerves) and enzymatic reactions in the body. It is a major coenzyme in many reactions in the body and is needed in the skeletal system. About seventy percent of ingested magnesium goes to bone growth, thirty percent to neuromuscular transmission (Pugh, 2002, p. 24; Coleby, 2006, p. 95). It is used in enzymes in the gut and muscle, helps prevent mastitis and acetonemia and promotes health of skin (Levy, 1991, p. 172; Coleby, 2006, p. 93).

Deficiencies

Deficiency of magnesium leads to grass, lactation and travel tetany, nervousness (especially apparent in horses) and increased risk for mastitis. Inflammation is a sign of deficiency and oxidative damage increases. Magnesium binds to receptor sites in the body where toxic metals might bind, deficiency in magnesium allows heavy metals to accumulate in the body (NRC, 2005, p. 224).

In horses, there can be sweating, paddling, ataxia and death from deficiency. Laminitis in horses is associated with magnesium deficiency, acute cases can be treated with Epsom salts. This should not be continued after initial treatment, however, since Epsom salts can eventually cause damage to kidneys. Chronic cases of laminitis can be cured with addition of a quality source of magnesium to the diet. Quick-growing forage in spring is often deficient in minerals, including magnesium, and this increases risk for laminitis. Some soils do not contain high levels of magnesium, sandy soils in particular, where leaching of magnesium can occur.

Horses that are shown and under stress will deplete magnesium reserves and have an acute case of laminitis after the event, illness or stressor. To prevent this, make sure a mineral is available that contains the proper ratio of calcium to magnesium (2:1) and consider addition of chelated minerals during times of stress. Herbs and homeopathy can be used to help reduce inflammation and heal laminitis in chronic cases (See Volume 2).

Other signs of deficiency in herbivores include uneven bone growth. Calcium deficiency diseases are related to magnesium deficiency. Urinary and renal calculi can occur, although this also may occur with toxic levels. Again, ratio with related minerals is critical for health, and proper water intake can help prevent some problems with magnesium imbalance. Deficiencies interfere with calcium absorption and the related vitamins and minerals (Coleby, 2006, p. 95; Pugh, 2002, p. 24; Abrams, 2000, p. 70-71; NRC, 1985, p. 13).

Toxicities

Magnesium toxicity is uncommon. When magnesium is taken in orally, even in high levels, toxicity does not occur unless the kidneys are not functioning well (NRC, 2005, p. 226). Soils in some areas of the United States can be high in magnesium and supplemental feeding of high magnesium continuously can throw off the balance with related vitamins and minerals, lead to urinary and renal calculi, diarrhea and weight loss (NRC, 2005, p. 227). The formation of the calculi in the urinary system can be avoided if there is enough fresh water available (not that this is an excuse for feeding minerals out of balance) (NRC, 2005, p. 227). If magnesium is force fed in very high amounts, used as an enema or injected, diarrhea and lack of appetite can also occur and extreme excess can have a sedating effect. Inability to stand can occur with toxicity, followed by coma and death (NRC, 2005, p. 226).

Ruminants

Ruminants are more susceptible to magnesium deficiency than other herbivores. Although ruminants make good use of poor quality grazing material, they are unable to digest minerals as easily and may become deficient in several minerals on pastureland that might support other herbivores adequately. Part of this has to do with how ruminants absorb minerals and the interference of other minerals with absorption of magnesium from forage (NRC, 2005, p. 224). Ruminants absorb magnesium from the rumen rather than the small intestine, although young animals in which the rumen is not yet fully functional absorb magnesium from the small intestine. Their needs for magnesium can be up to three times greater than the needs of non-ruminant herbivores and other animals (NRC, 2005, p. 225; NRC, 2001, p. 128). As was noted above, colder weather and damp, fast-growing forage tend to be magnesium deficient. Rumen pH also controls absorption, higher pH (over 6.5) decreases magnesium solubility. Conversely, grain causes the rumen to increase in acidity (decreasing pH) and magnesium is actually more available for absorption. Salt is important in magnesium absorption, without sodium, magnesium is removed by the kidneys. Potassium, on the other hand, decreases magnesium availability (NRC, 2001, pp.128-129). Salt should always be available for livestock free choice. Aluminum also interferes with absorption of magnesium (NRC, 2001, p. 129).

Pigs

Need for supplemental magnesium in pigs depends on stage of growth and diet. Deficiency symptoms include muscle twitching, irritability, weak pasterns and eventual tetany and death. Toxic levels have not been established (NRC, 1998, p. 50).

Horses

Horse mineral products tend to be low in magnesium, especially in relation to calcium. Although magnesium oxide, magnesium carbonate and magnesium sulfate are fairly well-absorbed (seventy percent) by horses, it is possible the supplements provide too little, depending on what else is available in the diet (NRC, 1989, p. 13). Deficiency symptoms are similar in horses to those in other animals. Nervousness, twitching, sweating, convulsions and death can occur. Low magnesium also leads to calcium and phosphorus deposits in blood vessels (NRC, 1989, p. 14). Because horses absorb most minerals better than ruminants, they can tolerate pastures low enough in magnesium to induce deficiency in ruminants. Stress increases the need for magnesium and horses that are shown, trailered or otherwise stressed are at risk for tetany, just like ruminants (NRC, 1989, p. 14). Toxic levels of magnesium have not been established for horses, dietary magnesium appears to be quite well tolerated (NRC, 1989, p. 14).

Carnivores

Ratio is as important in carnivores as it is in other species when considering magnesium and related minerals. Magnesium in proper ratio to calcium and phosphorus causes no concern for dogs but excess magnesium leads to seizures and problems with sodium and potassium transport in the body. Deficiency leads to lowered weight, vomiting, hyperextension in the front legs of puppies, convulsions, deposits in the aorta and blood vessels, problems with sodium and potassium transport and eventual death. Adults are more tolerant of deficiency than young animals (NRC, 1985, p. 18).

Plant sources

Pasture plants where soil is adequate in magnesium should contain enough magnesium to meet needs (unless needs are increased due to stresses or life stages). Leafy greens have been the recommended source of dietary magnesium for humans but again, soil content and life stages and stress play a role in the amount of magnesium available and how much is needed for health.

Phosphorus (P)

Phosphorus as an inorganic substance is the sixth most abundant mineral in the body and used in all parts of metabolism. Organic phosphorus is very toxic and used, unfortunately, as insecticides or herbicides where it does untold damage to the environment.

Phosphorus forms the matrix in bones that allow other minerals, like calcium, to be deposited. Deficiency of phosphorus, like calcium deficiency, leads to rickets (NRC, 2005, p. 290). Phosphorus and calcium are interrelated and almost all of both taken in are used in the skeleton. Fourteen to seventeen percent of the skeleton can be phosphorus and it is important in energy reactions in the body (ATP and ADP-adenosine triphosphate and adenosine diphosphate, respectively). Phosphorus is also used in the body in phospholipids, nucleic acids and phosphoproteins and helps maintain the acid/base balance in the body. Ruminants need phosphorus for proper rumen function, and levels needed depend in a large part on stage of growth and availability of calcium and vitamin D as well (NRC, 2005, p. 290). Phosphorus is also used in milk production, lactation increases needs (Pugh, 2002, p. 23).

Studies have shown that calves can meet their requirements for growth on grass-based feedlot diet without supplemental phosphorus (NRC, 2005, p. 290). Phosphorus has been over-supplemented due to the previously low cost of the mineral. Since it is eliminated in feces in high amounts (and some in the urine), this overfeeding can contribute to environmental concerns. In ruminants, current research and thinking is that supplementing phosphorus is unnecessary (Fluharty, p. 1). In pigs, the addition of phytase enzyme to the diet compensates for the pigs' inability to easily convert plant-based phosphorus (phytate) to a useable form. Ruminants, unlike simple stomach herbivores, have a greater kidney threshold for phosphorus reabsorption and elimination, which also allows them to subsist on diets of lower quality and less phosphorus. Increasing amounts from the diet lead to increased secretion from the kidneys (McDowell, 2000, p. 109).

Phosphorus is available as phytate from plants (although this form may not be as well absorbed in species like horses or pigs), phosphate in soils as dicalcium phosphate or bone meal (a poor source for herbivores!) (Pugh, 2002, p. 23; NRC, 1989, p. 11). The small intestine is the site where most phosphorus is absorbed by both active and passive transport, although ponies have been shown to absorb phosphorus from the colon. This absorption reaction requires sodium and vitamin D3 (NRC, 2005, p. 291).

The ratio of phosphorus to calcium is important. A ratio of 1:1 is usually adequate, depending on life stage, although some studies show that calcium can be up to seven times the amount of phosphorus without causing health problems (Fluharty, p. 1). When animals are fed high amounts of grain, there can be an imbalance in the phosphorus ratios.

In ruminants, phosphorus is secreted by the salivary glands, in part to act as a buffer for the volatile fatty acids in the rumen but also to help control the amount of phosphorus in the gut (and therefore absorbed by the small intestine). This absorption and secretion of phosphorus (and also calcium) appears to be unaffected by differing availability from the diet (McDowell, 2000, p. 108).

Homeostasis of phosphorus is controlled by the kidneys, intestines and bones. Three hormones control the reactions necessary to release, absorb or excrete phosphorus: calcitonin, parathyroid hormone and vitamin D3. In ruminants, the above-mentioned salivary excretion of phosphorus is what controls the phosphorus in the body, for the most part this occurs independently of any dietary phosphorus or lack. Older ruminants seem able to maintain phosphorus homeostasis more easily independent of dietary phosphorus (NRC, 2005, p. 292).

Because animals can maintain health with wide variations in their dietary phosphorus, especially if the phosphorus is in proper ratio to calcium, toxicity is rare from dietary sources. Excess phosphorus is detrimental due to the problems associated with urinary and kidney stones, interference with absorption of other minerals and hyperparathyroidism (fibrous osteodystrophy) in horses. This disease occurs when deposits of calcium are put into the connective tissues rather than into bones and can lead to large heads in horses ("big head disease") (NRC, 2005, p. 293). High phosphorus in cattle leads to lowered milk production, milk fever and hypocalcemia (NRC,

2005, p. 294). Pigs exhibit similar problems with very high levels or phosphorus and in birds, egg shell integrity and egg production is compromised (NRC, 2005, p. 294).

Deficiency

In deficiencies and imbalances, there can be decalcification of bones, rickets and loss of appetite (Abrams, 2000, p. 69). Symptoms are the same as for calcium and vitamin D deficiencies. Other symptoms include cystic ovaries, hyperparathyroidism (big head), enlarged upper and lower jaw and facial crest in horses. Blood levels are a better indicator than for calcium deficiency (NRC, 1989, p. 12)

Some authors consider deficiency rare in areas where phosphate fertilizer use is common but super phosphate fertilizers can throw off the balance of related vitamins and minerals and lead to deficiency symptoms of phosphorus, as well as the other minerals (Coleby, 2006, p. 106-107). Superphosphate fertilizer locks up magnesium, sulfur, copper, selenium, cobalt, boron, zinc and sometimes even phosphorus itself (Coleby, 2006, p. 106).

Because the body cannot free calcium from the bones without also taking an equal amount of phosphorus, any time the body needs more calcium than is provided by the diet, there is also the risk of phosphorus being removed from bone structures and released into the bloodstream (Abrams, 2000, p. 70).

Toxicities

As is the case with many minerals, toxicity symptoms are similar to deficiency symptoms. Fragile bones, elimination of trace minerals from the body and cystic ovaries are all related to toxicity (Coleby, 2006, p. 105).

Ruminants

Phosphorus, as noted above, has a potential to be an environmental toxin, especially when fed to ruminants. Providing enough for nutritional needs without furthering pasture and water contamination is the aim. Since phosphorus is needed in more biological functions in the body than any other mineral, people tend to assume it should be provided in some form in order to meet nutritional needs. Estimating how phosphorus is used and absorbed and in what forms is difficult in ruminants because of the recycling effect of phosphorus excretion in the saliva. Most studies done on phosphorus do not accurately account for this. Add to this the importance of correct ratios with calcium, magnesium, vitamin D and other trace minerals, and differing needs with stages of growth, and the equation becomes complex indeed. The more phosphorus that is available in the diet, the less the ruminants absorb and the less efficient is their salivary absorption of phosphorus. Unlike simple stomach herbivores, ruminants make good use of the phytate form of phosphorus, therefore alfalfa and pasture provide enough phosphorus for most animals to maintain themselves and reproduce (NRC, 2001, pp. 110-114).

Pigs

Phosphorus is always considered in its ratio to calcium and vitamin D is needed for absorption of either mineral. In pigs, the concern is less about phosphorus than it is about the form phosphorus is in. Unlike ruminants, pigs cannot make use of phytate unless the enzyme phytase is also available. For more information on phosphorus, see Minerals, subheading Calcium Pigs.

Horses

If there is one word that repeats throughout any discussion of calcium, magnesium and phosphorus in diets, it is "ratio." Mammals usually require a ratio of calcium to phosphorus of 1:1 or close, although some animals tolerate wide variation in that ratio and stage of growth of the animal may change the ratio slightly. Horses are no exception to this ratio and like pigs, phytate forms of phosphorus are difficult for them to absorb. Horses do have phytase enzyme in their lower digestive tract and this may increase digestibility of phytate (NRC, 1989, p. 11).

Deficiency causes similar changes in the body to all mammals, problems in bones and calcium availability lead to rickets and hyperparathyroidism. In horses, this manifests as the "big head" syndrome mentioned above.

Carnivores

Vitamin D, calcium and phosphorus are just as important in carnivores as in herbivores, for all the same reasons. The ratio is also important and the phytate form of phosphorus is much less useful to carnivores than to any herbivore. Deficiency of calcium leads to hyperparathyroidism and osteoporosis. Jaws are affected first and then other bones. Convulsions and tetany can occur with calcium deficiency, although rickets is not as common in adult dogs as in puppies. These symptoms would also relate closely to phosphorus and vitamin D deficiency as well (NRC, 1985, p. 16).

Sodium (Na)

Sodium, potassium and chloride, like magnesium, calcium and phosphorus, are interrelated in their uses in the body. While this book divides each ion into a different section, in reality they cannot be so easily separated. Sodium is almost always found in the environment as sodium chloride (salt) and potassium as potassium chloride. In this book salt is assumed to be sodium chloride and the terms are interchangeable. Because sodium, chloride and potassium are all related in their uses in the body and availability, excess of sodium will depress potassium (Coleby, 2006, p. 109). And excess potassium can lead to magnesium and calcium deficiency (NRC, 2007, p. 123).

Salt is absorbed straight through the small intestine wall (up to ninety-five percent of the ingested salt is absorbed this way) but is also secreted through the small intestine walls (above where it is absorbed), in the saliva, bile, gastric and pancreatic juices. Failure to reabsorb the secreted sodium chloride in the small intestine causes diarrhea that can lead to electrolyte loss and death (NRC, 2005, p. 358).

Kidneys are the site where sodium and chloride function to control water balance in the body in conjunction with hormones like aldosterone and vasopressin (NRC, 2005, p. 358). See Chapter 12 Organs of Detoxification and Elimination for more information on kidney function.

The chloride ion, in addition to being necessary in cellular function and kidney function, is used in hydrochloric acid. Chloride is critically important as part of the hydrochloride production as stomach acid. Hormones control the excretion of sodium and secretion of chloride. Usually sodium and chloride are either both excreted or both secreted, but in order for the digestive system to maintain a low pH (acidic environment), chloride must be present and sodium excreted from the system (NRC, 2005, p. 358).

Animals easily develop a craving for salt and will travel many miles (if given access) to find soils high in salt. Initially, salt craving causes animals to eat odd things trying to satisfy their craving, but eventually, the imbalances the deficiency causes leads to anorexia, poor hair coat, lack of milk production and death. Chickens become dehydrated during deficiency (NRC, 2005, p. 357). Detecting this deficiency by testing is not easy since the body will maintain normal blood levels as long as possible.

Forage is generally not high in salt (unless near oceans) but does contain chloride. Addition of salt as blocks or loose salt is essential for livestock in most areas (NRC, 2005, p. 357). Horses may be able to adapt to lower salt in the diet than ruminants, although horses will consume more salt than they necessarily need, while ruminants will self-limit their salt intake (NRC, 1989, p. 13). This fact allows mineral manufacturers to put salt into mineral formulas and blocks in order to limit the intake of the mineral in the block because the ruminant will stop eating when enough salt has been consumed. The downside is that if the trace mineral in the mix is needed in higher amounts, the ruminant is not likely to consume enough. See Chapter 7 for more information on mineral supplements and salt.

Salt is used in the body for extracellular cations and as part of electrolytes. Muscles and nerve function require salt and salt is in blood and even in bones (NRC, 2005, p. 357). Water metabolism in the body depends on salt, as do intracellular and extracellular functions and acid-base balance (Pugh, 2002, p. 23).

Salt is an alkalizer, strengthening digestive juices, aiding in iron assimilation, reducing premature hardening of tissues, preventing catarrh (mucus) and diseases of mucus membranes, maintaining health of the urinary system (Levy, 1991, p. 172-173).

Deficiencies

Deficiency of salt may lead to retarded growth, reduced fertility, impaired ability to use digested protein and carbohydrates and possibly decreased milk production (Abrams, 2000, p. 72). Loss of appetite and decreased drinking occur with low salt intake. During acute deficiency, muscle contractions and chewing become uncoordinated and blood potassium levels increase. Other possible symptoms related to salt deficiency include cancer, potassium deficiency and degenerative conditions. Since salt is important in digestion, lack will mean improper digestion and weight loss. There can be decreased skin rigidity although this can be difficult to determine (Underwood, 1981, p. 61).

Horses will lick objects where salt accumulates (like handles of tools) in order to try and meet needs (NRC, 1989, p. 13; Coleby, 2010, pp. 34-35).

Toxicities

Just as with any other mineral, salt is needed in the correct amounts. Excess salt contributes to water retention and edema, although true toxicity is unusual due to the body's ability to excrete excesses (Abrams, 2000, p. 73). Salt will also replace potassium in the body, causing other problems (Coleby, 2006, p. 109).

Too much salt can kill! It is my belief that adding salt to supplements and minerals increases the risk that animals will over-consume salt to meet their other mineral and nutritional needs, or conversely, end up over-consuming other minerals in an attempt to meet salt needs. Excess salt can also come from artificial fertilizers and poor irrigation practices (Coleby, 2006, p. 109-110).

Other problems associated with excess salt include cancer, degenerative diseases and prevention of proper digestion and assimilation of food (Coleby, 2006, p. 109-110).

Salt uses, needs, deficiency and toxicity symptoms are similar for most all herbivores and pigs. Slight differences are noted above.

Carnivores

Dogs and carnivores in general have a low dietary sodium requirement. Natural diets can meet this need or the addition of small amounts of organic kelp can be used. Slightly more than minimum requirement of salt increases water intake but does not appear to adversely effect health (NRC, 1985, p. 17).

Potassium (K)

Potassium is the third most abundant mineral in the body, but must be in balance with sodium and magnesium (Coleby, 2006, p. 109). Neither sodium nor potassium are stored in the body so they both must be taken in daily (Abrams, 2000, p. 72). Potassium is usually high enough in harvested forage to meet needs but is lower in grains and stored forages (NRC, 1985, p. 14). Fast-growing plants are often higher in potassium than the grazing animals require (Ammerman, Baker & Lewis, 1995, p. 295). Because of this, many nutritionists do not make note of potassium in the diet, it is assumed animals take in more than enough in forage. High-energy diets for maximum growth or lactation, however, can cause problems with potassium and deficiency may become a concern (Underwood, 1981, p. 59). Cool season grasses and legumes are higher in potassium than warm season species but older, mature pastures that are dried out, or hay that has been rained on, may have significantly less potassium than needed by ruminants (NRC, 2005, p. 308). Potassium can be inhibited by iron, low pH and "poor" soils (Coleby, 2006, p. 108).

Potassium is important in the acid/base balance in the body, in intracellular ions in muscles and skin, in nerve impulses and osmotic pressure and is important in enzymatic pathways (Pugh, 2002, p. 24; NRC, 1985, p. 14). Sodium occurs in blood plasma but not in cells, while potassium occurs in the cells where it often works with phosphorus in many metabolic reactions (Underwood, 1981, p. 61). Phosphorylation of creatinine and carbohydrate and protein metabolism require potassium, as do heart and kidney tissue. Insulin and other compounds help regulate the distribution of potassium in the body and cellular energy and gradients help move potassium and sodium into, and out of, cells (NRC, 2005, p. 307).

Absorption takes place in the small intestine, omasum and rumen (in ruminants) and excess potassium is excreted from the kidneys through urine (Ammerman, Baker & Lewis, 1995, p. 295; NRC, 2005, p. 307). Digestive problems can lead to lowered absorption, as they can for many other vitamins and minerals (NRC, 2005, p. 307). If sodium is deficient in the body, under the influence of the hormone aldosterone (and the adrenal glands), potassium begins to replace sodium in saliva. This is an adaptive strategy for times of low salt availability in the diet and should be viewed as such an indicator

(Underwood, 1981, p. 61). It is not a sign of health and illustrates the importance of providing salt as a free choice supplement.

Ruminants require more potassium than non-ruminants and the needs of each animal species depend not only on species differences but on stage of growth and other factors, like stress. Like other vitamins and minerals involved in antioxidant and stress management, potassium needs and use increases during heat or travel stress, and with amount of protein in the diet (NRC, 2005, p. 306).

Deficiencies

Deficiency symptoms include decreased feed intake (this will also occur with disease, cobalt and sulfur imbalances, among others), decreased live weight gain, listlessness, stiffness, impaired response to sudden disturbance, convulsions and death (NRC, 1985, p. 14). Blood vessels become constricted (and in pregnancy and labor, this contributes to inability of fetuses to turn head down for proper birth presentation), termed dystocia, or difficult birth. Deficiency of potassium and vitamin C at conception interferes with the true pattern of inheritance. This means it is imperative that breeding animals have proper mineral supplementation in order to pass on the best of their genetic potential. There will be detectably lower amounts of potassium in the blood during deficiency (Coleby, 2006, p. 108). Navicular disease in horses is related to potassium, as is repetitive strain injury in humans (Coleby, 2006, p.108).

Toxicities

Excess potassium creates magnesium depression, decreased energy and decreased weight gain, and could cause cardiac arrest (NRC, 1985, p. 15; 1989, p. 12). Grass tetany, the usually sudden occurrence of tetany in young grazing animals on lush spring grass is often not explainable by low magnesium in the diet but the excess potassium in early, fast-growing pasture plants that can throw off the balance of magnesium and induce sudden cases of tetany. These cases do respond to a single, high dose of magnesium. This is more likely to occur on pastures fertilized with potassium (or potash) but can occur even without fertilizer (Underwood, 1981, p. 54). Otherwise, toxicity is rare because the body generally eliminates any excess. Kidney failure is the main cause of potassium build up in the body but other physical problems can contribute, including low insulin, certain medications, genetic differences, acidosis and cellular damage (NRC, 2005, p. 311). The form of potassium fed in supplementally high amounts makes some difference as to possible toxic conditions. Potassium bicarbonate in excess leads to cancers and imbalances in the pH in the body while potassium chloride in excess leads only to excessive amounts of K+ ions rather than other deleterious changes in tissues (NRC, 2005, p. 309). Excess of potassium and sodium in diets of dairy cows has shown to increase incidence of milk fever (NRC, 2005, p. 309).

Ruminants

Because potassium is not stored in the body, it is required daily. This requirement, however, is easily met under most conditions by grazing and diet. Exceptions include continuously feeding hay that has been stored long periods or put up after being rained on, or left to dry too long. Other concerns include any time the digestive system is not functioning properly or the animal is under stress. Diets that are high in grain and low in forage also contribute to deficiency, since grains are usually low in potassium (NRC, 2007, p. 123). Kelp is a source of potassium (and other minerals and vitamins) and can be used as part of a balanced mineral supplement for all livestock and animals (Thorvin, 2013).

Deficiency symptoms are typical of all animals: poor growth, decreased feed intake, stiffness, muscle paralysis (NRC, 2007, p. 123).

Toxicity can occur when pastures have been fertilized with potassium or in spring when grazing fast-growing grasses, but usually symptoms are not severe. Extreme and sudden toxicity can lead to hypomagnesia (grass tetany) or milk fever (low calcium). Since molasses also contains high levels of potassium, continuous feeding of molasses in higher amounts can also contribute to problems. This is in addition to the concern about excess sweet carbohydrates that lead to systemic fungal infections and increased attractiveness to parasites (NRC, 2007, p. 123).

Pigs

Like other animals, pigs require potassium for many reactions in body tissues. Potassium is the third most abundant mineral in the body. But normal diets do provide adequate potassium under most circumstances. Deficiency symptoms are the same as for other animal species. Toxicity is rare, pigs tolerate very high amounts of potassium as long as they have fresh water available (NRC, 1998, p. 51).

Horses

Grazing provides the needed potassium for horses under normal conditions like it does for ruminants. Deficiency is uncommon, as is toxicity (NRC, 1989, p. 12).

Carnivores

Potassium is provided in a raw, real food diet by the tissues from herbivores that contain adequate potassium. Deficiency induces similar symptoms as it does in herbivores: poor appetite and growth, muscle paralysis, dehydration and lesions on the heart and kidneys. Commercial diets for dogs are often high in salt and potassium with apparently no deleterious effects (NRC, 1985, p. 17). This is taken to mean that toxicity is not usually of concern, not that commercial processed diets are healthy!

Chlorine (Cl)

Chlorine is an intracellular ion used to maintain osmotic balance and is a component of gastric secretions (HCl, stomach acid) (Pugh, 2002, p. 23; NRC, 1985, p. 11). It contributes to the suppleness of joints and tendons, removes toxic elements from tissues, prevents over-formation of fatty tissues and promotes health of teeth and hair (Levy, 1991, p. 170).

For more information, see the section on Sodium above.

Sulfur (S)

Sulfur is found as amino acids in feed and forage or added to mixes. The methionine form is best absorbed, then sulfate, with the least absorbable being elemental sulfur (NRC, 1985, p. 15)

Sulfur is used in wool, keratin, skin and the nervous system. It is contained in proteins and used for making sulfur-containing amino acids and other compounds, such as cysteine, cystine, taurine, homocysteine, cysteic acid, cystathionine, methionine, glutathione (important as an antioxidant with selenium), fibrinogen and some estrogens (NRC, 2005, pp. 28-29). Sulfur is also essential in the synthesis of the B vitamins biotin and thiamine and other compounds like heparin, insulin and chondroitin sulfate (Abrams, 2000, p. 74; Pugh, 2002, p. 24; NRC, 1985, p. 15).

The body can convert methionine to several sulfur-containing compounds but it cannot convert inorganic sulfur to methionine. In ruminants, rumen bacteria need sulfur for proper function and to synthesize methionine and B vitamins. This means ruminants need inorganic sulfur in fairly high amounts. Non-ruminants, on the other hand, cannot synthesize some of these compounds from sulfur and have no inorganic sulfur requirement. Instead, these animals need the particular compounds as part of their diet. For example, cats cannot synthesize taurine from methionine and must have taurine as part of their diet (NRC, 2005, p. 373).

During digestion and absorption, dietary sulfur is converted to proteins and sulfur-containing compounds that are absorbed through the intestinal walls. Body cells use amino acids that contain sulfur as their sulfur source, excess sulfur is eliminated in the urine (NRC, 2005, pp. 373-374).

For herbivores, grazing provides sulfur in amino acid forms from plants. Alfalfa contains a decent amount of sulfur and herbivore diets with enough protein usually provide enough sulfur. For carnivores, meat and fishmeal provide sulfur (NRC, 2005, p. 374).

The form sulfur is in makes a large difference in how available it is for digestion and, therefore, how potentially toxic. Obviously, forms like hydrogen sulfide gas are toxic and can build up in confinement operations where animals, pigs in particular, are kept over manure pits. Rumen bacteria in ruminants can also produce this compound that leads to polioencephalomalacia. Uncooked brassica family plants (broccoli, kale, etc.) lead to low thyroid function and eventual hemolytic anemia (NRC, 2005, p. 374).

In animals, some otherwise common and unconcerning products can become problematic. Dimethyl sulfoxide (DMSO) is a compound added to topical products for inflammation in horses (in particular). This sulfur-containing compound readily crosses the skin and carries with it any other drugs or compounds in the product. In one case, horses were poisoned by mercury that was a carrier in a product containing DMSO (NRC, 2005, p. 375).

Sulfur as elemental sulfur is not very absorbable and is not particularly toxic. Large amounts of sulfur in a single dose can be toxic to the point of death. Flowers of sulfur has been used in the past as an internal (and external) treatment for various problems in animals, including injuries and parasites. Internally in large doses, this can be fatal. Adding sulfur in higher amounts to the diet over time does not initially cause problems other than diarrhea (NRC, 2005, p. 375).

Deficiencies

Since sulfur is part of the amino-acid complexes (proteins), deficiency involves loss of appetite, reduced wool growth and reduced weight gain. Other symptoms of deficiency include excess salivation, tearing from the eyes (lacrimation) and possibly external parasites (lice). See also Cobalt for other causes of weight loss and lacrimation. Rumen bacteria must have sulfur for proper function, so deficiency in sulfur leads to improper rumen function and all the symptoms associated with that (NRC, 1985, p. 15; Abrams, 2000, p. 74; Coleby, 2006, p. 112). Because sulfur is part of the B vitamin complexes, deficiency will lead to inability of ruminants to synthesize these vitamins. Non-ruminant deficiency symptoms can be different and will be addressed below under the different species.

Toxicities

One of the concerns with excess sulfur is the problem associated with other mineral interactions. Sulfur forms compounds with copper and molybdenum, decreasing their usage and also decreases selenium retention. This has not been well-studied in equines but appears to occur in ruminants (NRC, 1989, pp. 14-15; Coleby, 2010, p. 36). High levels of sulfur result in decreased rumen function, impaction and colic, jaundice, odor of hydrogen sulfur on the breath, anorexia and eventual death (NRC, 1985, p. 15). In addition to sulfur's role as a compound with molybdenum and copper, sulfur by itself can interfere with copper absorption, the organic methionine form of sulfur is just as problematic as the inorganic (Underwood, 1981, p. 105).

Ruminants

Toxicity of sulfur in ruminants leads to central nervous system problems, blindness and seizures. Note that these symptoms can be caused by other conditions as well. These are different than symptoms in non-ruminants, where diarrhea is one of the more common symptoms. In ruminants, there can be brain lesions (also indicative of several other problems) in conjunction with the polioencephalomalacia (NRC, 2005, p. 376). The rumen bacteria convert excess sulfur to sulfide that is then burped up during cud-chewing and absorbed through the lungs. The absorption of the hydrogen sulfide through the lungs is the problem and this is the cause of the polioencephalomalacia (NRC, 2005, p. 376). Rumen bacteria can adapt to higher sulfur diets over time, leading to no symptoms of polioencephalomalacia or other problems. In general, polioencephalomalacia is considered the same as thiamine deficiency, but addition of thiamine in sulfur-toxic ruminants does not correct the neurological problems (NRC, 2005, p. 377).

Other issues seen with high sulfur include interference with trace minerals and reduced feed intake. Extra sulfur can come from various sources, including feeds and well water so some care must be taken to determine what might be out of balance on the farm (NRC, 2005, p. 377).

Sulfur forms a complex with molybdenum and copper that can make copper completely unavailable for digestion. Sulfur also interferes with selenium absorption (NRC, 2005, p. 377). See below under Trace Minerals for more information.

Pigs

Pigs may benefit more from addition of amino acid forms of sulfur rather than inorganic sulfur (NRC, 1998, p. 51).

Horses

Deficiency is not well understood in horses. Extreme toxicity can cause colic, lethargy, yellow frothy discharge from the nose and jaundice. Convulsions and death can follow (NRC, 1989, p. 15). The interaction between copper and sulfur in ruminants may not occur in horses (NRC, 1989, p. 15).

Carnivores

Methionine is the form of sulfur needed by carnivores and this should be adequate in a real food diet containing enough natural protein sources (NRC, 1985, p. 11). Addition of excess methionine to the diet can cause toxicity in cats, with hemolytic anemia and Heinz-body formation as symptoms (NRC, 2005, p. 376). Adding particular supplemental nutrients to a commercial diet for carnivores can lead to imbalances and certainly does not provide the trace nutrients found in real, whole foods. This is yet one more reason to avoid commercial diets and instead focus on feeding carnivores as close to nature as possible.

CHAPTER 10

MICRO OR TRACE MINERALS

Trace minerals in this book include but are not limited to selenium, copper, cobalt, molybdenum, manganese, iron, iodine, zinc and boron. These minerals, unlike the macro minerals, are not needed in high amounts, but they are absolutely critical for health and life in all animals and people. They are also needed in proper ratio and balance to one another, just as the macro minerals like calcium, phosphorus and magnesium are needed in balance with one another. Imbalances can create deficiencies and toxicities that interfere with all aspects of health, from immunity to reproduction and growth.

Because our soils are now depleted, plants have no access to trace minerals for uptake, and herbivores become deficient on otherwise healthy-looking pastures. Contrary to what some practitioners would have you believe, plants cannot contain trace nutrients that are not available in soils, although it is entirely possible we are unable to test for all the trace minerals that are available at some level to plants. Carnivores fed on meat from deficient herbivores do not gain the trace minerals they also need for health. Commercial fertilizers increase the imbalances rather than working to correct them, and our farming practices do not take into account local ecosystems and breed differences. Trace mineral imbalances have become the new health epidemic in animals and people. There are so many things that could be said here about the problems associated with modern farming practices and the disassociation from the earth that nurtures life and the God that gives it. Our livestock have become fragile creatures no longer able to even subsist, let alone thrive, on the earth when once they were the bringers of life for humankind. They have been manipulated through artificial breeding practices and poor management decisions, their very DNA has been changed due to poor nutrition so they no longer produce progeny that better the breed and become a source of pride to the farmer. Animals are managed as commodities that are overcrowded on small, unnatural lots and fed nutrient-depleted feeds in processed forms. These poor beasts never have an opportunity to realize the once-great potential of their ancestors and will be lucky if they see the pasture that is their due. As a source of food, they are not any more nutritious for the people or animals that eat the meat than the manipulated plants grown in depleted soils are a source of nutrition for these herbivores. The circle of life is waning.

The following minerals are in a list as separate minerals, like the macro minerals. Like the minerals above, however, these minerals cannot really be separated from each other and the nutrients they form, and interact with, in the body. The first three, selenium, copper and cobalt, are given first because these are the three minerals that are so often depleted. Imbalances in these three minerals cause untold numbers of health problems that are often credited to diseases and parasites rather than their true cause: mineral imbalances.

Selenium (Se)

Problems with selenium in livestock have a long history but the first understanding of selenium was from its ability to build to toxic levels. Records from 1857 relate symptoms of toxicity in horses (Ammerman, Baker & Lewis, 1995, p. 303). In the modern world, however, deficiency is the more common condition (Ammerman, Baker & Lewis, 1995, p. 303). The common forms of selenium in soils are selenite, selenate, elemental selenium and the form available from plants, selenomethionine (NRC, 2005, p. 321). Because not all soils contain selenium, there are parts of the U.S. and the world that are extremely selenium deficient. This can contribute to a host of problems and make some pastures unable to support herbivores, particularly ruminants. Deficiency is a common concern in many areas, coupled with low vitamin E (See Chapter 8). Conversely, breeds of livestock that have developed in areas where higher selenium levels are available naturally may tolerate or need higher levels of this, and other, trace minerals.

Bio-available complexes of methionine (selenomethionine and selenocysteine) are used in supplements to supply easily-absorbed forms (NRC, 2005, p. 321). The selenocystine, cysteine and methionine forms are found in forages and grains grown on selenium-rich soils (Underwood, 1981, p. 149). The form of selenium ingested contributes to availability in the body. Selenate is better absorbed than selenite and the organic forms are very well absorbed by all species (NRC, 2005, p. 323). Methionine forms are more available, therefore accumulate in tissues more quickly. Alltech, maker of Selplex organic selenium, has done studies confirming that there appears to be no LD50 for Selplex (telephone conversation with Alltech representative). The LD50 is based on a test used to determine toxicity of a substance. The amount of the substance given to cause fifty percent of the test subjects to die is the LD50. Inability to set an LD50 means that in the study, the researchers were unable to feed enough selenomethionine to kill half the test subjects.

Elemental selenium is not well absorbed, making it less toxic but also less useful for supplementation in low selenium areas. Selenides are less toxic than selenites or selenates. Selenium in seleniferous feeds is more toxic than soluble salt forms (Underwood, 1981, p. 162).

Selenium can interact with iron or sulfur, making it unavailable and contributing to deficiency (Pitzen, 1993, p. 9). Certain plants, known as converter or accumulator plants, uptake selenium at a greater rate, causing toxicity in animals ingesting them, although these plants are not necessary for toxicity to occur. *Astragalus bisulcatus* (milk vetch) is one such plant, although its toxicity may be more related to alkaloids than selenium levels (Merck 2011). Coal and municipal waste combustion release selenium into the atmosphere but there is, so far, no evidence this fallout effects levels in soils (Underwood, 1981, p. 160). The toxic waste that is released into air, soil and water from coal power plants, however, does cause toxicity in increasing amounts up the food chain.

While this book emphasizes the importance of proper mineral supplementation in livestock, the health problems associated with mineral deficiencies in humans are becoming more prevalent and better understood. In the case of selenium, there are several chronic diseases related to deficiency, including a cardiomyopathy and a disease that causes malformations in the joints of the hands (osteoarthropathy) (NRC, 2005, p. 322). Selenium-deficient dietary muscular dystrophy is a debilitating disease of weakening muscles. Progression and mortality from AIDS may relate to low selenium. Several studies have been done that confirm people with AIDS have lower levels of selenium than those without the disease and that supplementation in this and other chronic diseases can reduce symptoms (Dworkin, 1994; Bella, et al., 2010). For people who eat meat, the health of the animals that will eventually be food is also of concern, and proper mineral supplementation means people get needed trace minerals from food animals as well.

Selenium is part of glutathione peroxidase, an antioxidant that protects cells from damage by reducing lipids to alcohols and peroxides to water. Peroxides are damaging and reactive substances that destroy cell membranes and inhibit normal steroid production in the body (NRC, 2001, p. 195). Selenium occurs as selenoproteins (including selenocysteine) and is needed for prevention of white muscle disease as well as cancer (NRC, 2005, p. 322). One protein form includes the iodothyronine 5'-deiodinase, an enzyme needed to change thyroid hormones from one form to another (T4 to T3, for instance (NRC, 1998, p. 55).

Absorption takes place in the small intestine and can be inhibited by other minerals, like sulfur, calcium and lead, or by the plant alfalfa. The kidneys and liver are storage sites for selenium in the body, blood plasma levels fall rapidly after supplementation (NRC, 2005, p. 323). Selenium is removed from the body primarily by urine but some is also removed in feces. Selenomethionine and selenocysteine supplementation result in higher levels in all relevant body tissues, in animals and also in milk. These organic forms of selenium cross the placenta as well, resulting in young born healthy with plenty of needed selenium (NRC, 2005, p. 323).

Absorbed efficiently in non-ruminants, selenium is needed for growth, fertility (especially in males for sperm) and oxidative protection of red blood cells (erythrocytes). Selenium is also important for prevention of deficiency diseases. Horses absorb up to seventy-seven percent of available selenium (NRC, 1989, p. 17). Selenium and vitamin E work synergistically, each nutrient helps offset deficiency from lack of the other. Both are important in antioxidant and immune system functioning (Jain, 1993, p. 215; McDowell, 2000, p. 169).

Selenium is poorly absorbed in ruminants, as little as twenty-nine percent of selenium taken in may be available for digestion and nutrition (NRC, 2001, p. 142). This depends on the form of selenium available, chelated forms are much more readily absorbed and can be used to offset deficiency or, to some extent, interaction with other minerals.

Selenium appears to work synergistically with copper. When enough copper is available in the diet, the levels of selenium needed in order to be toxic increase (i.e. copper can decrease the toxicity of selenium). Conversely, feeding selenium to animals on marginal copper diets improved copper levels (NRC, 1985, p. 21). Selenium and vitamin E work together, without vitamin E, the body cannot use selenium. Forage, however, should contain adequate levels of vitamin E. Stored hay may have lower levels, and supplementation of vitamin E may be necessary in winter. Both vitamin E and selenium function as antioxidants in the body (Underwood, 1981, pp. 149-150). See Chapter 8 for more information on vitamin E.

Deficiencies

Deficiency of selenium leads to white muscle disease (dietary muscular dystrophy), infertility, unthriftiness, weak lambs at birth or shortly thereafter (NRC, 1985, pp. 20-21; Underwood, 1981, pp. 149-150). Unlike some minerals, serum, plasma or blood levels can be useful measurements to determine deficiency and toxicity. It is important to remember that, in deficiency, blood levels can drop quickly after supplementation so any testing should be done before supplementation or several days after.

Vitamin E and sulfur are important in utilizing selenium. If feeding vitamin E improves the condition, selenium is at least part of the underlying deficiency. Pasture and hay should provide plenty of vitamin E unless the hay has been stored long enough that vitamin degradation has occurred. See Chapter 8 on Vitamins for more information on the interactions between selenium and vitamin E.

In white muscle disease, skeletal and cardiac muscles are affected. There can be acute, subacute or chronic forms signified by weakness, impaired movement, difficulty nursing in newborns, respiratory distress and heart conditions. Blood tests show elevated creatinine kinase, aspartate, aminotransferase, potassium and urea

(NRC, 1989, p. 17; Coleby, 2010, p. 34).

Not only is selenium poorly absorbed but several other minerals interact with selenium and interfere with uptake by plants. Sulfur, gypsum or superphosphate fertilizers decrease selenium uptake by plants and may contribute to selenium deficiency (Underwood, 1981, p. 163).

In newborn animals, deficiency is easy to see as it leads to weak, cold lambs and kids, unable to stand to nurse. Placentas do not pass quickly (they should pass within an hour or so after birth) and the weak young will often die without intervention. In the wild, herbivore young must be able to run alongside their mothers at, or very shortly after, birth to avoid predators. Domestic livestock should still be able to run very shortly after birth, young should be standing and walking within a minute of birth if not moving before their feet hit the ground! In severely selenium-deficient newborns, even the tongue muscle can be affected and the young may be unable to suckle. Supplementation at this point may help if the young is also tube-fed and kept warm. But the easier way to do this with a better outcome is to prevent the problem to begin with. If supplementation is necessary, consider a selenomethionine form, even gelcaps made for humans that also contain vitamin E. These supplements can be given daily, but again, this is time-consuming and unseen internal damage from deficiency in utero may result in less-than-satisfactory growth and health in the developing youngster. Sudden death may even occur once the young animal is up and running because the exercise may be too much strain on a damaged heart muscle. Prevention is the key here and supplementation should be done all year.

It is worth noting that other mineral deficiencies can result in weak newborns, low copper causes problems with the nervous system leading to "swayback" disease in lambs. Selenium deficiency is by far the more common concern over most of North America and in many other countries. As our management practices become more and more extreme, deficiencies of trace minerals will become more common.

Toxicities

Toxicity can take several forms and occurs mostly when animals are either in a starvation situation and choose to eat unpalatable accumulator plants, or ingest enough high selenium soils during normal grazing to induce toxicity. In subacute forms of toxicity, the animals grazing selenium-rich pastures get "blind staggers." This manifests generally as animals stumbling around, wandering with problems in vision. Front legs become weak and all symptoms increase in severity with time and exposure. In extreme cases, the tongue becomes paralyzed and the animal eventually dies. In more chronic cases, the symptoms are referred to as "alkali disease." Chronic selenium toxicity (alkali disease) is not usually the result of accumulator plants but can be from low-level, long-term exposure, either from pasture or supplementation. Alkali disease results in loss of wool, lack of vitality, emaciation, stiff joints and erosion of joints and atrophy of the heart. Remember that deficiency may produce heart defects also. Sloughing of hooves can occur, as well as reduction in reproduction and possibly death (NRC, 2005, pp. 326-327). Severity can vary greatly. According to "Mineral Tolerance of Animals" (2005), sheep grazed on pastures moderately high in selenium for several months have shown only slight wool loss on the neck and goats given selenite daily in doses up to one mg/kg body weight (BW) for almost a year showed no symptoms. Increasing selenium to five mg/kg did cause death (p. 327). To mitigate the toxic effects, feeding low-selenium forage can allow the body to eliminate the excess (NRC, 1985, p. 22; Underwood, 1981, pp.160-161).

Injectable selenium and vitamin E has caused death, but probably from anaphylactic shock due to carrier ingredient in most cases, unless repeated injections are given over a long period (NRC, 1989, p. 19). Many farmers are now becoming aware of the importance of selenium and supplementing. Just because a little is good, however, does not mean a lot is better. Excessive amounts of supplemental selenium can result in toxicity and many of the symptoms listed above. Keeping minerals in balance with one another helps prevent toxicity and interference in absorption between minerals.

Ruminants

Studies on dairy cows have shown that retained placenta is associated with lower levels of antioxidants in the body (NRC, 2001, p. 195). Supplementation of both vitamin E and selenium are needed to correct this deficiency and stop the occurrence of retained placentas (NRC, 2001, p. 196). Remember also that other vitamins and minerals play an antioxidant role and indeed, deficiencies of vitamin A and iodine also contribute to retained placentas in dairy cattle (NRC, 2001, p. 196).

Llamas and alpacas generally handle normal dietary levels of selenium quite well but can easily become deficient in areas where soils and forage are deficient. Supplementation may be necessary but over-supplementation can cause the same concerns of toxicity as it can in any other animal species. Using a bioavailable form of selenium as part of a balanced mineral supplement is the best option for maintaining good health in camelids. Be prepared to read labels, though. Many supplements marketed for camelids contain minerals out of balance and in forms not well absorbed. These should be avoided just as overfeeding and supplemental grain should be avoided. Camelids in North America tend to be very overweight with teeth that actually need to be cut to keep them from growing into the upper palate. This is ridiculous since these animals have survived for millennia in South America without having their teeth cut. Natural grazing on grass and browse wears down teeth and provides nutrients without the excess weight gain or need for dental procedures.

Pigs

Studies have shown that selenium levels in pigs may be higher when the form of selenium includes sodium selenite, not just the bioavailable forms (like selenomethionine) (NRC, 1998, p. 56). The selenite form appears to accumulate to toxic levels quicker than the methionine form (NRC, 2005, p. 327). Phosphorus in the diet interacts with selenium but calcium has less, to no, impact on selenium needs (NRC, 1998, p. 55). In pigs, glutathione peroxidase is an indicator of selenium status, low selenium means less glutathione peroxidase. Selenium deficiency in pigs can often result in sudden death, although this is less common in other animal species where deficiency symptoms occur before death, giving farmers the chance to remedy the problem (NRC, 1998, p. 56). Toxicity does not occur at normal levels but higher levels can cause the same types of toxicity symptoms seen in other livestock, including hair and weight loss, degeneration in the kidneys and liver and hooves sloughing (NRC, 1998, p. 56). In pigs, there can also be problems of degeneration in the spine in toxicity from *Astragalus bisulcatus* that may result more from a toxin in the plant (swainsonine) than the excess selenium (NRC, 2005, p. 328).

Horses

Deficiency is similar in horses to symptoms in other animals and is linked to vitamin E levels. Horses will be weak with problems in swallowing, breathing and heart conditions (NRC, 1989, p. 18). Acute toxicity symptoms include blind staggers, blindness, head pressing, perspiration, abdominal pain, colic, diarrhea, increase heart rate and respiration and lethargy. Chronic selenium toxicity is called alkali disease and includes symptoms of alopecia, especially of the mane and tail, and cracking hooves around the coronary band. Hooves may be deformed or slough off and this condition causes the animal to be unwilling to move. Problems in the liver can occur as well as kidney and heart problems (NRC, 1989, p. 18; NRC, 2005, p. 327).

Carnivores

Selenium is important in carnivores for the same reasons as it is needed in herbivores but the diet of meat should provide enough selenium, assuming the food animals were fed a selenium- and vitamin E-adequate diet (NRC, 1985, p. 21).

Plant sources

This is not an all-inclusive list but some sources of selenium include Brazil nut (*Bertholletia excelsa*), catnip (*Nepeta cataria*), milk thistle plant (*Silybum marianum*), quack grass (*Elytrigia repens*), valerian root (*Valeriana officinalis*), blue cohosh root (*Caulophyllum thalictroides*), barberry root (*Berberis vulgaris*), blessed thistle (*Cnicus benedictus*), marshmallow root (*Althaea officinalis*) (Walters, 2013, p. 257).

Copper (Cu)

Throughout antiquity, copper has been used as a precious metal, aid to healing and valuable addition to alloys. It is possible that the mining of copper may have resulted in trade between North America and Phoenicia in ancient times. Copper's ability to kill invertebrates accounts for its use as a treatment in soils and water for snails and slugs and reduction of algae. It is also used as a treatment for its antifungal properties (NRC, 2005, p. 134). There are environmentally safer methods of reducing populations of unwanted mollusks or algae, however. And susceptibility to fungal infections is an indication of internal imbalance with a diet too high in refined sugars and carbohydrates. Nutritionally, copper is often touted as a cure for arthritis, even when used externally in jewelry. Too many people claim relief from minor (and sometimes major) arthritic pain for this practice to be dismissed as quackery. Before the advent of plastic PVC plumbing (which may yet prove to be as big a health hazard as lead paint), copper piping possibly provided too much copper in many households and can still contribute to contamination of samples, as can plastic and some paints.

Copper is available commercially as copper sulfate, copper oxide or copper carbonate. Of these, copper oxide is the least available for digestion, in fact, it is almost completely unavailable. The inability of animals to make use of copper oxide makes it a perfect choice for copper oxide wire particle (COWP) boluses used in sheep to reduce parasites.

Only one major parasite is affected by copper outright and that is barberpole (*Haemonchus contortus*). This parasite is of great importance to ruminant producers because it attaches to the abomasum walls and sucks blood, leaving the host anemic to the point of death. If an animal is maintaining high levels of this (or any) parasite, the animal has been unable to build up an immunity to the parasite and this relates to all aspects of health and nutrition. Parasites are a symptom, as much as a cause, and should be viewed as a clue that the animal is not in good health or the animal would not maintain such detrimentally high levels of parasites. Unfortunately, indiscriminate breeding, lack of pasture management and poor nutrition now contribute to high parasite loads in many flocks and herds.

See below for more information on sheep and copper, but it is erroneous to think that sheep cannot have copper. Sheep must have copper but how much depends on several factors and can be difficult to estimate. The COWPs do not provide copper for digestion to sheep and can be used safely in any breed; however, the COWPs are not addressing the issue of parasites at its root, and if the need for COWPs arises, the management and nutrition of the flock or herd should be re-evaluated.

One concern about copper availability involves copper and water. Copper in some forms becomes less available in water. Copper sulfate is one of the only forms of copper that does not become less available once it is suspended in water. Soil that contains copper may not provide nutritional amounts because copper reacts with other volatile sulfides in soil to form precipitates (NRC, 2005, p. 136). Chelated forms of copper are becoming more common. Copper sulfate is not palatable and is caustic to mucus membranes, and copper oxide and copper carbonate are not at all available for absorption. For supplementation, copper chelate is a good choice if soils and forage are deficient. Grasses are higher in molybdenum and lower in copper, legumes are the reverse (Underwood, 1981, p. 106; Coleby, 2006, p. 100). Forage, however, may not provide useable copper to ruminants, even if the forage contains higher levels of copper, due to interactions with other minerals (molybdenum and sulfur being two common culprits) (NRC, 2005, p. 137).

Molybdenum, sulfur and iron interact with copper to a greater or lesser extent (Arthington, 2003, pp. 11-12; Pugh, 2002, p. 24; Underwood, 1981, p. 91). During drought, plants uptake molybdenum at higher rates, further depleting copper in the system. Excess zinc also interferes with copper absorption (Ammerman, Baker & Lewis, 1995, p. 136).

Ruminants and non-ruminants digest copper very differently and their need for copper can be drastically different because of this. Ruminants tend to make poor use of copper, and their absorption of copper compared to non-ruminants is less than ten percent (NRC, 2005, p. 135). This inability to absorb copper is related to interaction of other minerals in the digestive system, like molybdenum, sulfur and zinc. Young ruminants whose rumen is not yet fully functional actually absorb copper much better because much of the interactions between minerals occurs in the rumen (NRC, 2005, p. 135).

Absorbed copper can be stored in the liver or eliminated from the body via bile and feces. In some species, genetic liver problems can lead to accumulated copper in the body. In people, Wilson's disease causes excess copper. The excretion of copper in bile is the main mechanism for copper removal in non-ruminants but in ruminants this may not be as efficient and in sheep, higher dietary copper does not cause an increase in excretion in the bile. This leads to accumulation and eventual toxicity (see below under Copper Sheep) (NRC, 2005, p. 135).

Copper in plasma is bound to a protein called ceruloplasmin. This protein carries both copper and iron from one part of the body to another via the blood stream. Stored copper in the liver is bound to metallothionein, which also binds to other metal minerals and heavy metals. This process allows for detoxification of the excess metals. Sheep do not increase metallothionein production much when copper is ingested in higher amounts, resulting in more copper problems (NRC, 2005, p. 135).

Copper is used in connective tissues and as part of the myelin sheaths of nerve fibers. It allows the body to convert absorbed iron into hemoglobin and supports the integrity of the mitochondria (the source of energy for cellular function). There are enzyme reactions that depend on copper. Proper heart function also relies on adequate copper. Deficiency can lead to heart failure and death. Note that selenium deficiency can also lead to problems with the heart muscle that lead to death. Copper is also used in melanin synthesis, allows detoxification of superoxide, is important in immune health and resistance to parasites and fungal diseases. Proper ovulation depends on adequate copper, as does proper bone growth, healthy wool and proper wool color (NRC, 2005, p. 134). This becomes obvious in black sheep that will silver out in copper deficiency. In fact, many years ago, shepherds kept a black sheep or two in the white flock to see if copper deficiency had become a problem. If the black sheep began to silver, the shepherd knew the whole flock needed more copper.

In other species, copper also affects color of hair but the effects can be more subtle. Black goats may have sunburned tips to their hair, horses may bleach out and colored sheep may show signs of sunburning in the wool, colors not as dark or wool being of wrong texture. The color differences alone should not be used as the only measure of copper deficiency since colored sheep may have variation in the fleece that is the result of genetics or other factors. The black fleeces silvering, however, can be striking and worth noting.

In cattle, there have been recorded instances of spontaneous bone fractures from weak bone structure due to copper deficiency. Young animals may be born with deformity in bone and cartilage or may develop heart defects later (around six months of age (Abrams, 2000, p. 76; Coleby, 2006, pp. 99-100; NRC, 1989, p. 15; Coleby, 2010, p. 25-26).

Deficiencies

In many parts of the world, deficiency of copper in grazing animals is much more common than either normal levels or toxicity of copper (Ammerman, Baker & Lewis, 1995, p. 128). Deficiency of copper is very often related to other antagonistic minerals. Molybdenum or sulfur (or both) in the diet in ruminants is directly related to the amount of copper available for absorption. Excess of either molybdenum or sulfur will lead to copper deficiency. Lack of adequate sulfur in the diet, however, can lead to toxicity (especially in sheep) (NRC, 2005, p. 135). Sulfur forms indigestible copper sulfide in the digestive system. Molybdenum interferes with copper absorption as well and molybdenum and sulfur form thiomolybdates that can also form indigestible compounds with copper in the rumen (NRC, 2005, p. 136; Ammerman, Baker & Lewis, 1995, p. 136).

In non-ruminant herbivores, copper does not interact in the same way with molybdenum but can interact with excess amino acids, like methionine and cysteine (NRC, 2005, p. 136). Zinc in excess can cause problems in both ruminants and non-ruminants (NRC, 2005, p. 136).

While chelated copper is a good source of bioavailable copper in most cases, in areas where excess molybdenum causes problems, addition of some copper sulfate to a mineral supplementation may be advantageous. The theory is that copper sulfate will form complexes with the molybdenum and sulfur, leaving the copper chelate for absorption and health.

Copper deficiency leads to anemia, although this is not the only cause of anemia (Underwood, 1981, p. 93). Iron without copper is not useable for red blood cell production, which is important to remember when considering anemia from *Haemonchus contortus* infection. Ceruloplasmin causes iron to be absorbed from the intestines, transported to the tissues and used for hemoglobin synthesis (Underwood, 1981, p. 93). Iron can inhibit vitamin E, so some authors recommend avoiding iron tonics or injections. Quite often the limiting factor in anemia is not the iron, it is copper. Since anemia can also relate to cobalt deficiency (pernicious anemia), that is worth investigating. See below under Cobalt for more information (Coleby, 2006, p. 100). If iron is the limiting factor, a food source of iron is a better option than injections or concentrated supplements. Stinging nettles (*Urtica* spp.) and dandelion (*Taraxacum officinale*) are two common weedy plants that contain decent amounts of bioavailable iron.

Neonatal ataxia is associated with copper deficiency in the pregnant ewe. "Swayback" is muscular incoordination and partial paralysis of hindquarters that results from degeneration of myelin sheaths of nerve fibers. There can be progressive ascending paralysis, lack of nursing and death. Deficiency can result in weak lambs either at birth or a short time after birth (Underwood, 1981, p. 94). This can be difficult to distinguish from selenium deficiency but the weakness results more in nerve problems and paralysis rather than muscle weakness. The best prevention for this is good mineral supplementation in the pregnant ewe. If a lamb must be treated for selenium or copper deficiency, using a chelated source is best and quickest. Copper deficiency is better treated by supplementing the ewe, the damage to the lamb is already done and takes time to correct. Nervous system support may help and supplemental tube feeding will be necessary if you expect to save the lamb. Here is another example where proper nutrition will prevent a host of problems in newborn and developing young.

Steely or stringy wool with little elasticity, silvering fleeces (on dark animals), wool that lacks tensile strength and affinity for dyes are all indications of copper deficiency. Hair color depigmentation is a sign of copper deficiency in all animals, although, as mentioned above, how this manifests can differ slightly between species (Underwood, 1981, p. 95). Camelids do not show the same color changes in wool during deficiency as do other species.

Bone disorders, such as osteoporosis or spontaneous bone fractures in adults, occur during deficiency. Infertility and scouring (diarrhea) can also occur. The scouring can be due to high molybdenum that forms complexes with copper and leads to deficiency (Underwood, 1981, pp. 93, 98).

Animals that are copper deficient have increased susceptibility to parasites, diseases and skin infections. They develop cardiac lesions leading to "falling disease" or spontaneous death. Liver concentrations will be low, therefore body tissue concentrations should be taken. Plasma should be tested in particular, since blood levels may be falsely elevated due to stress (Abrams, 2000, p. 76; NRC, 1985, p. 17; Coleby, 2006, pp. 99-101; Underwood, 1981, pp. 92-96; Pugh, 2002, pp. 24-25).

Other diseases associated with copper deficiency (and therefore lowered immunity) are cancer, foot rot, cowpox, ringworm, foot scald, proud flesh, Johne's (Crohn's in humans) and brucellosis. These diseases are also associated with low iodine, manganese and cobalt (Coleby, 2010, p. 25-26).

Animals deficient in copper (and possibly other trace minerals) may chew fences and bark. Since trees bring up nutrients from the subsoil that may be otherwise unavailable, it is possible animals are trying to meet their nutritional needs by eating bark.

Zinc suppresses copper; however, the reverse is not true (Ammerman, Baker & Lewis, 1995, p. 136). It is important that minerals be in balance with one another for good health. Copper interacts with molybdenum, sulfur, zinc, selenium, silver, cadmium, iron and lead, in addition to other minerals occurring out of balance (NRC, 1989, p. 15). In fact, lead toxicity leading to anemia is lessened when there is adequate copper in the diet. The mechanism for this is presumed to be because lead interferes with copper in iron metabolism. Enough copper can offset this detrimental interaction (Ammerman, Baker & Lewis, 1995, p. 136).

Another potential problem in copper absorption is the use of high levels of ascorbic acid (vitamin C) (Ammerman, Baker & Lewis, 1995, p. 137). Ascorbic acid interferes with copper absorption and, under normal circumstances, this can be problematic if someone chooses to use vitamin C continuously at high doses. In copper toxicity, however, this reaction can be put to good use. Vitamin C should be part of the therapy for copper toxic animals (usually sheep).

Toxicities

In sheep, the margin of error between deficiency and toxicity is smaller than in other livestock and much of this relates to the sheep's inability to remove copper via bile using the metallothionein protein. The availability of antagonistic minerals in the diet makes evaluating the amount of copper needed by sheep very difficult. Sheep, in some cases, can easily tolerate high levels of copper because of the presence of the interacting minerals. Without the presence of those other minerals, however, some sheep breeds will become toxic on the same levels of copper that are tolerated well by sheep in the presence of antagonistic minerals. Lambs are more at risk for toxicity since the interactions with other minerals happen in the fully functioning rumen, something lambs do not have. Absorption is increased in lambs and toxicity is more an issue (NRC, 2005, p. 138). Horses and cattle tolerate copper well and toxicity is rarely an issue. Goats and llamas have similar copper requirements and, in general, do well with copper.

It is possible that the burden of our chemical farming methods on the livers in animals (particularly sheep) may contribute to the intolerance of levels of trace minerals that in years past were considered adequate for true health. Restoring the health of the farm and animals can improve their ability to use the trace minerals available and to eliminate those not needed, rather than building up toxic levels.

Different breeds have differing susceptibilities: angora goats can be very susceptible while Nubian goats are less susceptible (Radostits, et al., 2000, p. 1601).

Sheep and Copper

As has been stated several times, in sheep, the margin of error between deficiency and toxicity is smaller than in other livestock. This really cannot be overstated. Goats, llamas, alpacas and other livestock can tolerate much higher levels of copper than most sheep breeds (Pugh, 2002, p. 25; Fowler, 2010, p. 31). But sheep do need copper and other factors may interfere with their copper absorption and availability, leading to deficiency. Care should be taken to evaluate soils, forage and sheep use of copper to determine if extra supplementation is necessary. Different breeds of sheep have different requirements for copper. Some European breeds may not need much copper, coming from coastal areas that are low copper. Other breeds, like Icelandic sheep, may require copper amounts similar to those fed to goats.

It is worth noting copper toxicity symptoms in sheep. These include increased respiration, depression, weakness, hemoglobinuria (hemoglobin in urine causing red color), icterus (jaundice) and death. A veterinarian can help you determine cause of death in suspected copper toxic sheep by doing a necropsy, checking liver, kidneys and blood levels of copper (Pugh, 2002, p. 25).

As mentioned above, different breeds have differing susceptibilities: Ronaldsay, Orkney and Friesian sheep are more susceptible. Merino sheep and dwarf goats may require more copper than other breeds (Pugh, 2002, p. 25).

If it is determined that a sheep has copper toxicity (the red urine and jaundice are good clues if no other testing is available), there are a few things that can be done to counteract the excess copper. High levels of vitamin C can be used since copper toxicity inhibits vitamin C production in the body and because vitamin C in high doses inhibits copper absorption (Ammerman, Baker & Lewis, 1995, p. 137). Vitamin C in high levels is also used to treat other poisonings, primarily due to vitamin C's strong antioxidant properties. It is worth keeping injectable vitamin C or ascorbic acid on hand. In a pinch, rose hips can be beneficial but how much vitamin C is in the hips depends on species of rose, when and where it was harvested, processed and stored. Dolomitic limestone has been used as an adjunct therapy to the vitamin C and activated charcoal can help with any copper still in the digestive tract. Since excess copper damages the liver and kidneys, supporting these organs is important. Milk thistle seed powder (*Silybum marianum*) is useful in liver disease, toxicity and failure and may help protect both liver and kidneys. Some homeopathic remedies can be useful. Homeopathic Apis is used for any organ that is swollen and may help support the liver and kidneys. Other liver and kidney detoxifying remedies would be appropriate but remember that the copper stored in the liver is not the problem, causing release of more copper into the bloodstream may exacerbate the situation. Any detoxifying should be done slowly and after the acute poisoning is under control (Coleby, 2006, p. 101). Veterinarians use molybdates to bind up copper and this has been shown to be very effective in toxic sheep (NRC, 2005, p. 136).

It is important to note that part of the toxicity of copper in sheep comes as the copper is released from the liver. I have heard some people mention herbs that help detoxify the liver as appropriate for sheep with copper toxicity. My personal opinion is that these herbs are contraindicated and may in fact cause more problems and symptoms. Herbs like burdock root (*Arctium* spp.), dandelion root (*Taraxacum officinale*), yellow dock root (*Rumex crispus*) and others with similar properties would, in this author's opinion, be best avoided in acute toxicity. Milk thistle seed (*Silybum marianum*) works differently and has been shown to help regenerate liver tissue. It is sometimes used in hepatitis and cirrhosis for this reason and would be a better choice for supporting liver function in toxic sheep.

Ruminants

Much pertinent information on copper requirements in ruminants is covered in the sections above. This is a good place to elaborate on the metallothionein protein that binds to copper for transport. This protein also transports zinc and heavy metals and is important in detoxification for this reason. Once the liver reaches the upper limit of its ability to store this protein attached to metals (like copper and others), under stressful situations these metals (and copper!) are released into the bloodstream. This is what causes toxicity in sheep and can occur at 1000 mg copper/kg DM (dry matter) (NRC, 2007, p. 127). How much dietary copper it takes to reach this level depends on other factors (see above). In goats, copper needs are similar to sheep and so are requirements; however, increased levels of copper in the diet may help immunity, milk production and growth. Meat goats may require more copper as do lactating (dairy) animals (NRC, 2007, p. 128).

In addition to the swayback disease that occurs in copper deficient newborns, there can also be a delayed reaction to deficiency that produces ataxia and nerve damage at about one to two months of age (NRC, 2007, p. 129).

Toxicity in ruminants (particularly sheep) has two parts: the toxicity that occurs when the liver is accumulating high levels (toward that 1000 mg copper/kg DM) and once the copper is released into the bloodstream, causing hemolytic anemia, jaundice and loss of blood in the kidneys. Liver enzymes become elevated and the liver becomes damaged. The first part of the toxicity equation can last for several weeks and the second, acute stage lasts for hours or days (NRC, 2007, p. 129). When using copper with sheep, it is imperative that the farmer pays close attention to health of the animal and is prepared to treat toxicity, if necessary. Remember that the amount of copper that is absorbed and accumulates in the animal is very dependent on more than just the amount of copper in a supplement. Weather, mineral interactions and life stages of the animals are all important and can change rapidly. This means otherwise healthy animals can show toxicity when least expected.

Goats, both meat and dairy, have a much higher tolerance of copper than any breed of sheep. While there have been established upper limit levels for dietary copper in sheep (assuming normal levels of molybdenum and sulfur), as of the writing of the Nutrient Requirements of Small Ruminants (2007), there had not been an upper limit established for goats (p. 129).

Pigs

Copper has been shown to promote growth in pigs and poultry, which has not escaped the notice of commercial hog farms and industry. Swine feeds are often high in copper, so much so as to be toxic to sheep (and some cattle species) if manure and waste from pigs is applied to fields, then used as ruminant grazing or hay ground (NRC, 2005, pp. 134, 139). All of this is not to say that pigs require high copper but that their tolerance of it, and resulting weight gain, become an advantage to commercial farm operations.

Horses

Copper is used for the same metabolic reactions in horses as it is in ruminants but horses not only absorb the non-chelated copper well, they also tolerate higher levels of copper than ruminants. Copper can interact with several other minerals (see above) but these interactions may be less important in horse nutrition because they may not occur during digestion like they might in the rumen of ruminants (NRC, 1989, p. 15). Deficiency leads to some of the same symptoms as in other species, including bone problems, but toxicity appears to not be of much concern (NRC, 1989, p. 15).

Carnivores

The role copper plays in hemoglobin formation is important in carnivores as well as herbivores. Deficiency causes anemia that cannot be reversed without copper (NRC, 1985, p. 19). Copper requirements in dogs is not high and toxicity can occur if copper is supplemented (NRC, 1985, p. 19).

Cobalt (Co)

See also Chapter 8 Vitamins under B12 for more information

Cobalt is available as cobalt sulfate, cobalt oxide, cobalt carbonate and several chelated products including cobalt glucoheptonate. For more information on uses, see below but note that injections of cobalt are ineffective in preventing deficiency symptoms related to rumen bacterial die-off and subsequent nutritional deficiency. Ruminants can pull cobalt from the liver to make vitamin B12 but cannot use cobalt from liver stores to put back into the digestive system. Injections of vitamin B12 can be effective for preventing vitamin deficiency but are cost-prohibitive and time-consuming compared to feeding cobalt regularly, and injections do nothing to offset the bacterial die-off.

In most animal species, cobalt is needed for synthesis of vitamin B12, cobalt being the center of the vitamin. Both iron and cobalt are needed in order for the body to make vitamin B12 and vitamin B12 is the cofactor for several enzymes that support propionate and methionine synthesis. Methionine is important in folate reactions in the body (NRC, 2005, p. 124). Vitamin B12 is stored in the liver and kidneys. Animals can survive in cobalt-deficient areas for a few months off their bodies' reserves before depletion becomes apparent, and equines can tolerate lower levels of cobalt than ruminants, allowing them to thrive in areas where ruminants would die. Low levels of deficiency may not show anemia, only unthriftiness and weight loss (Underwood, 1981, p. 115).

Vitamin B12 is not available in forage but is made by bacteria in the digestive tract (the cecum and colon in horses). In humans choosing a vegan diet, supplementation of vitamin B12 is necessary.

Acidic soils (low pH) and artificial fertilizers may inhibit the availability of cobalt (NRC, 1989, p. 15, 25).

Non-ruminants have a similar transport system for iron and cobalt, iron deficiency allows cobalt to be absorbed to a greater extent. Cobalt is excreted in urine and stored in liver, kidneys and heart (NRC, 2005, p. 125).

In ruminants, cobalt is used by rumen bacteria in synthesis of vitamin B12. Vitamin B12, in turn, aids in recycling methionine and is therefore important in folate metabolism. Methionine is an essential sulfur-containing amino acid. Folate helps form red blood cells while reformation of methionine is from homocysteine (Underwood, 1981, p. 119). As noted above, although liver stores of cobalt can be used to prevent vitamin B12 deficiency even in times of low dietary intake, rumen bacteria do not have such stores and within days of lack of cobalt intake will switch from propionate to succinate production (NRC, 2001, p. 132). This affects nutrition and ruminants receiving too little cobalt will begin to show signs of anorexia and weight loss.

Because of the need rumen bacteria have for cobalt, ruminants need cobalt in addition to that required for manufacture of vitamin B12. When cattle were fed cobalt amounts higher than needed to meet B12 requirements, they had enhanced rumenal digestion, especially of lower quality forage. This may either be a result of selection of rumen bacteria which need more cobalt or links between bacteria and feeds that allow bacteria to cling to the feed particles better (NRC, 2001, p. 133). The results are taken to mean ruminants may have a higher requirement for cobalt than their vitamin B12 requirements would suggest.

It is also worth noting that for manufacture of red blood cells, the body cannot use iron without cobalt and copper (Walters, 2013, p. 120).

Since cobalt is synthesized into vitamin B12 in the digestive tract, stress, illness and drugs can hinder conversion. Be aware that using FAMANCHA may be misleading. Cobalt deficiency/pernicious anemia can be mistaken for heavy parasite infestation. Do fecals to verify your visual findings. Lack of eggs may indicate underlying deficiency, not parasites. Parasites do not lay eggs at all times of the year or stages of growth, lack of eggs may not indicate lack of parasites, either. And remember that anemia, from parasites or other causes, cannot be reversed with iron alone, there must also be copper, cobalt and related vitamins.

Deficiencies and Strategies for Correcting Them

Because cobalt is so important to proper rumen function in ruminants, I am including a list of values from trace mineral analysis from liver tissues. I have not done this with other minerals because values tend to be a wide range and interactions with other minerals mean each situation may be unique. Cobalt is crucial in ruminants so I have opted to include this information. Always check with each lab's recommended ranges for trace minerals, as techniques for evaluating liver results differ from lab to lab. Severe cobalt deficiency fresh liver values in ug/g: < 0.07, moderate deficiency: 0.07-0.10, mild deficiency: 0.11-0.19, correct range: > 0.19 (Underwood, 1981, p. 115).

Low cobalt will result in a lack of vitamin B12, lack of appetite, lethargy, emaciation, anemia and "wasting disease." White liver disease can occur and there may be ophthalmic discharge (tearing or rheumy eyes- see also Sulfur). Animals will be pale (from anemia) and may suffer from fatty liver, decreased estrus, and decreased milk and wool production (Abrams, 2000, p. 77; NRC, 1985, p. 19; Pugh, 2002, p. 25).

Deficiency of copper coinciding with cobalt deficiency makes determination more difficult. Cobalt-deficient animals will be colder than normal, although, of course they can also become ill and run a fever. If there is no other explanation and an animal is lethargic, has little appetite and is losing weight but is colder than normal, cobalt deficiency is likely (Coleby, 2006, p. 97; Abrams, 2000, p. 77). Cobalt and iron are absorbed through similar pathways in the body, during iron deficiency, cobalt absorption increases (NRC, 2005, p. 125).

Determining deficiency requires testing but observation can go a long way to deciding what needs changed in a mineral supplementation program. Looking at a sheep (or other ruminant) that is cobalt deficient you will see some obvious, and some not so obvious, signs. The animal may appear to be eating but on closer inspection, the animal is only snuffling through the hay or picking at grass blades and not eating mouthfuls of food with gusto. No matter how rich the food, the animal will almost certainly be thinner than it should be. Wool may be of poor quality and eventually slough off. The eyes may have a discharge that is either clear or may start to show signs of infection and look like pus or yellowish discharge. If infection is present in the eyes, treat accordingly. See Volume 2 on disease treatment for more information. Cobalt-deficient sheep (or goats) are going to be less able to digest rich feeds and grains and may be more at risk for bloat. They may also refuse grain or richer feeds in favor of more roughage, leaves and weeds. Since digestion is impaired, there can be pain and the animal may be grinding its teeth. This is easily mistaken for chewing cud, which the animal may not be doing often, or at all, further hampering proper nutrition. The animal begins to have a very unthrifty appearance and may look hunched with back legs swung under. As vitamin B12 deficiency progresses, the sheep will become weak and want to lay down often. Standing becomes more difficult and this can be mistaken for selenium deficiency. Since digestion is not adequate, other vitamin and mineral deficiencies creep in.

In order to correct this, several things need to be done. First, make sure to get enough cobalt into the diet and start feeding a cobalt-rich supplement to the afflicted animal. Since bacteria are dying in the digestive system, a source of probiotics helps considerably. If possible, grabbing a bit of cud from a healthy animal and feeding it to the cobalt deficient one can jump start rumenal digestion but this can be tricky. Molars on ruminants are very sharp and sticking your fingers in the animal's mouth is a great way to get bit. Yogurt can work but needs to be fed often. Human probiotics will work in a pinch and livestock probiotic products are available that can work very well. My only complaint about livestock probiotic products are the inert (and often unhealthy) ingredients.

The next thing that should be done is vitamin B complex injections that include vitamin B12. Since digestion is impaired, vitamins B complex are not being manufactured or absorbed correctly and all of these vitamins may be deficient, along with vitamin C. Oral supplementation is less effective since the animal is not digesting well. The B vitamins can also help boost appetite but without probiotics, increase in food intake still won't really help.

Now, providing food that is digestible becomes paramount and it is not as easy as it sounds. Like an ill person who has had intestinal flu, the sheep has poor digestion and cannot be expected to immediately benefit from, or digest, rich feeds. Instead, forage with a lot of minerals and fiber content is a good choice. Tree and bush leaves, raspberry and blackberry leaves, rose bushes and grasses with a lot of choices in forbs and species diversity are all beneficial. It goes without saying that you should never harvest these from areas that are sprayed or along roadways where exhaust accumulates on plants. If you have the option to put the sheep out to pasture in an area with varied grazing, this is perfect but remember that the sheep is weak. It may be unable to return to a water trough or shelter. It also may be at risk for predation because it is weak and falls behind the flock, away from livestock guard dogs and other guard animals.

All types of nutritional support are useful here, as is common sense in dealing with a weakened animal. Unfortunately, even the best of care may not be enough to correct a deficiency. The damage may be too severe, and in some cases, there appears to be an underlying genetic inability to tolerate even low-level deficiency. Studies done on sheep to determine deficiency levels of cobalt showed that during times of low-level deficiency, most animals muddled along. Some animals became ill and a few died (Stewart, 1950, p. 325). With proper supplementation, most animals recovered completely but some never did recover well and continued to decline. This seems to be the case in many flocks that have deficiency problems. Most animals will recover, some animals need extra care and may never recover fully and a few individuals will die, no matter what is done for them at that point. This is not a reflection on the shepherd! It is an unfortunately consequence of the basic physiology and genetics of our livestock. This illustrates the importance of providing an adequate amount of cobalt at all times to prevent even low-level deficiency. In a flock managed for production, animals that have had problems with deficiency should be culled as these may not recover completely and cannot go on to produce to the best of their ability. You also do not want to pass on genetics for animals unable to weather low-level deficiency. If you have a pet or fiber flock, obviously your focus is different and the above information can be used to help save compromised animals.

Why do I write so much about cobalt? I believe in ruminants that this is the base of the digestive pyramid. Without cobalt, other nutrients become unavailable due to poor digestion. Even good-quality feed is not enough to maintain the animals and poor quality feed is worthless to them. This does not need to occur! With proper cobalt supplementation, ruminants become unbelievably efficient at converting even low-quality forage into weight gain. A flock getting the cobalt it needs becomes an asset and is easier to maintain, tolerates parasites better and produces healthy offspring, good quality wool and plenty of milk. The flock costs less to raise and maintain if cobalt is given in the needed amounts.

How to Supplement Cobalt

Cobalt is not available in many plants, availability depends on soil acidity and the species of plant. Higher pH decreases cobalt uptake by plants. Legumes have more cobalt than grasses, assuming there is available cobalt in soils (NRC, 2005, p. 125). Animal sources can be low in cobalt as well, although liver is generally high enough in cobalt (NRC, 2005, p. 126).

Because of the way small ruminant digestion works, minerals like cobalt can easily filter down to the bottom of the layers of fibers in the rumen. From there, the cobalt is much less available to bacteria, and low-level deficiency symptoms start to occur. Cobalt sulfate is fairly well-used by ruminants but it still needs to be fed regularly and may filter down. Chelates and other forms of cobalt formulated to stay in suspension in the rumen longer can help mitigate this effect. Under normal conditions, cobalt propionate, cobalt carbonate and cobalt sulfate are all similarly bioavailable, but cobalt oxide is not well absorbed at all (NRC, 2005, p. 126).

Cobalt boluses can help somewhat but have some major drawbacks. The bolus tends to settle to the bottom, once again becoming less available to rumen bacteria, and can also become covered in deposits, meaning it cannot be properly digested and assimilated. Boluses and pellets can be regurgitated and spit out (NRC, 2001, p. 132). A loose mineral with several forms of cobalt can be a good choice for flocks and herds that have cobalt deficiency.

Toxicities

Cobalt toxicity is not usually considered problematic in ruminants or other animals (Pugh, 2002, p. 26; NRC, 2005, p. 126). Adult ruminants can handle much higher levels than required without any side effects although calves and lambs which do not yet have fully functioning rumens may be at risk for toxicity. Cobalt can pass with milk into the abomasum and from there, unchanged, into the intestines where it can cause problems. Injecting young animals with cobalt is also risky and does not solve the problem of inadequate cobalt in the diet. Offering a loose mineral with good levels of cobalt is a better option for supplying needed amounts without risk of toxicity (NRC, 1985, p. 19).

In cases where cobalt has been supplemented in extremely high amounts, symptoms of toxicity include lethargy, diarrhea, convulsions and death (NRC, 2005, p. 126). The levels required to produce toxicity when fed chronically differ by species and length of supplementation. Generally, animals with chronic toxicity have reduced weight gain but in ruminants, toxicity was tolerated. This may be due to other factors, like iron in the diet or supplemental methionine or cysteine. These amino acids can reduce toxicity as can selenium, vitamin E and vitamin C, which also help protect against organ damage by cobalt toxicity (NRC, 2005, pp. 126-127).

Since many herbivores end up as food for people and carnivores, it is noteworthy that cobalt tissue concentrations in herbivores fed high levels of cobalt are not high enough to cause toxicity in animals or people feeding on the tissues but kidney levels of cobalt may be above the toxic limit (NRC, 2005, p. 127). Organs can accumulate minerals to higher levels than most other body tissues and should be eaten or fed in lesser amounts. Not to mention the risk of heavy metal toxicity in these detoxifying organs as well.

Ruminants

Sheep are more susceptible to cobalt deficiency than cattle (Erickson, 2015). All ruminants use some of the cobalt in binding with rumen bacteria and this hinders cobalt absorption for vitamin B12 synthesis. In other words, ruminants need more cobalt that non-ruminants (NRC, 2005, p. 125). In the rumen, ruminants use cobalt to synthesize vitamin B12 but the rumen bacteria also synthesize other molecules from cobalt. These molecules resemble the B12 molecule but are not active; however, they can cause test results to show higher vitamin B12 than is actually present (NRC, 2001, p. 132).

Simple stomach non-ruminants are not as sensitive to cobalt deficiency as ruminants. This is due to several factors, including the need for vitamin B12 in gluconeogenesis (the conversion of non-carbohydrates to glucose) and reduction of propionic acid (NRC, 2001, p. 132). Ruminants can store vitamin B12 and cobalt in the liver, and this storage will last the animal several months during deficiency. The animal cannot remove the cobalt from the liver and return it to the rumen where vitamin B12 and cobalt are needed for proper microbial function and populations (NRC, 2001, p. 132).

Young animals are more sensitive to deficiency because they do not have liver stores built up. Deficiency signs in young and adult ruminants include poor growth, unthriftiness, anorexia, fatty degeneration in the liver, anemia and poor immune function (NRC, 2001, p. 132).

Pigs

In pigs, there is no need for cobalt in excess of what is needed by the body to make vitamin B12. Pigs can, however, use cobalt in place of zinc for several zinc-based enzymes. Toxicity is more an issue for pigs than ruminants and symptoms include anorexia, muscle tremors and anemia (similar to the deficiency symptoms seen in ruminants) (NRC, 1998, p. 52). Adequate vitamin E and selenium do help detoxify and protect against cobalt toxicity but excess copper (often used to increase growth rate in pigs) can increase the severity of the toxicity (NRC, 1998, p. 52).

Horses

Horses can thrive on pastures that are too cobalt deficient to support ruminants. Horse requirements, like those of pigs, are determined only by the body's need to make vitamin B12 and the B12 requirements are usually met by synthesis of the vitamin in the intestine (NRC, 1989, p. 29).

Carnivores

In carnivores, vitamin B12 is often supplied by the diet and cobalt becomes somewhat unnecessary (NRC, 1985, pp. 21-22). It is considered a trace mineral and in carnivores, as opposed to ruminants, the requirements are very low and not of particular concern in a natural diet.

Molybdenum (Mo)

Molybdenum occurs in all body tissues, as part of metalloenzymes (xanthine oxidase, aldehyde oxidase and sulfite oxidase), and is needed for fertility. Molybdenum is used by nitrogen-fixing bacteria and in enzymes involved in amino acid metabolism (NRC, 2005, p. 262).

There has been no evidence of molybdenum deficiency in sheep, apart from its relation to copper, and its addition to mineral formulas is due to the misguided assumption that sheep cannot have copper. The molybdenum is to offset any sources of copper that may be in the mineral or diet (NRC, 1985, p. 16; Underwood, 1981, p. 103). Like all trace minerals, molybdenum is needed in very small amounts but unlike some of the minerals previously discussed, molybdenum deficiency is rare in animals with access to a varied diet (NRC, 2005, p. 262). There have been cases of deficiency in goats where the diet was more purified (NRC, 2005, p. 262).

In ruminants, molybdenum is stored in the liver and excreted in urine or feces (depending largely on how much dietary sulfur is available). High sulfur and high copper to molybdenum ratios favor fecal excretion, as do high molybdenum and normal sulfur levels (NRC, 2005, p. 263). Molybdenum also crosses into the milk during lactation (NRC, 2005, p. 263). Unlike some of the other trace minerals, molybdenum levels in the body are better observed from plasma levels than from tissue levels (NRC, 2005, p. 263).

Non-ruminant herbivores use urine as the primary excretion for molybdenum. Interactions between copper, sulfur and molybdenum are not nearly as important in non-ruminants as they are in ruminants (NRC, 2005, p. 263).

Deficiencies

In animals fed purified or modified diets (to induce deficiency), deficiency symptoms include anemia and weight loss (NRC, 2005, p. 262). Molybdenum-containing enzymes help with the superoxide free radicals and trauma-induced inflammation in the body (that could be caused by parasites, for instance). In humans, there is at least one recorded case of a person with Crohn's disease who improved with molybdenum supplementation (NRC, 2005, p. 262).

In ruminants, there has, so far, been no evidence of deficiency under normal circumstances except as it relates to copper toxicity (Underwood, 1981, p. 103). Tungsten is an antagonist to molybdenum and can be used to induce (or can cause) deficiency (Underwood, 1981, p. 103). This is usually not an issue since tungsten is not common in the environment (NRC, 2005, p. 263). Because the transport system used by the body to absorb molybdenum in the small intestine and kidney is the same system used to transport sulfate, sulfur may be an antagonist as well (NRC, 2005, p. 263). Ruminants getting enough sulfur should not have a problem with excess molybdenum, although this can depend on other factors. Since ruminants do not normally have sulfur reach the small intestine (the site of molybdenum absorption) as sulfate, there may not be as much interaction as one might assume (NRC, 2005, p. 263; NRC, 2007, p. 133).

Toxicities

Some sources consider molybdenum toxic due to its ability to bind copper and lead to scouring, poor wool quality, anemia and death. These symptoms correspond to copper deficiency symptoms (Abrams, 2000, p. 81; Underwood, 1981, pp. 103-104). In areas where soils are high in molybdenum, it can be difficult to keep animals from suffering from copper deficiency and chelated copper sources are recommended. Molybdenum toxicity may be much less a concern in horses than in ruminants. Plants uptake more molybdenum from soils during drought stress so toxicity can vary widely by the year and weather conditions.

Ruminants

Ruminant requirements for molybdenum are low and differ by species, but in general, supplementation is not recommended (NRC, 2001, p. 141). Not much research has been done to establish requirements except as it concerns sheep and copper (NRC, 2007, pp. 133-134). Deficiency symptoms include lack of appetite and poor weight gain, reproduction problems and death (NRC, 2007, p. 134).

Horses

There is less concern about molybdenum interfering with copper in horses than in ruminants (NRC, 1989, p. 15).

Manganese (Mn)

Manganese does not naturally occur as an isolated metal but is part of over 100 other minerals (NRC, 2005, p. 235). Most of the manganese in the U.S. is used in making steel, added to gasoline in place of lead or as alloys in aluminum for soft drink cans. Other places manganese is used include batteries, animal feed or fertilizer (NRC, 2005, p. 235). Usually found in pasture soils, manganese content in forage can vary. Alfalfa content may vary from eight ppm to one hundred ppm. Soybean meal may contain thirty to fifty ppm (Underwood, 1981, pp. 130-131). Plants use manganese to help move iron into the chlorophyll molecule. Forms include manganese dioxide, manganous oxide and manganese sulfate and chelates. Chelated manganese can have an absorption rate of 125%, compared to manganese sulfate at thirty percent (NRC, 2001, p. 140).

Manganese is an activator for enzymes, including those needed for proper bone and cartilage formation and proper prothrombin production for blood clotting. It is used in reproduction, as well as normal bone development, and is distributed in cells and tissues (Underwood, 1981, p. 125; NRC, 2007, p. 132). Carbohydrate and fat metabolism require manganese and it is found in chondroitin sulfate, which is important in joint health.

Dietary manganese may not be very bioavailable, absorption that takes place in the gut can be as low as one percent and excretion in bile is very quick. Absorption depends on other mineral interactions, including calcium, iron and phosphorus. The amounts of phytate and amino acids also changes absorption rates (NRC, 2005, p. 236). From the gut, manganese is taken to the liver by transferrin, a glycoprotein containing iron. Storage depends on dietary intake, excretion is also in proportion to dietary intake. This means that increased amounts in the diet often lead to increased amounts being excreted, as well as stored, and levels in the body do not build up as easily as they can with some other minerals (NRC, 2007, p. 132).

Young born to mothers who had adequate manganese will not have deficiency, manganese crosses the placenta and is found in milk (NRC, 1998, p. 55).

Inhalation of manganese from soils, or other sources, like industrial pollution of the air, is not a usual way of acquiring manganese but it can cause toxicity when it does occur (NRC, 2005, p. 236).

Vegetables and forage are the main sources of manganese for people and animals but as was mentioned above, forage amounts can vary widely. Corn is not a good source of manganese and corn-based diets for birds or animals will lead to deficiency unless supplemental manganese is provided (NRC, 2005, p. 237).

Soy is also not a good addition to the diet for manganese and real milk rather than soy milk replacers leads to higher manganese uptake (NRC, 2005, p. 237). See Chapter 11 for information on real milk replacers and the importance of colostrum to newborns. Chelated manganese products increase availability, sometimes as much as one and a quarter times the absorption of manganese from manganese sulfate or chloride (NRC, 2005, p. 237).

Phosphorus and calcium can interfere with manganese absorption. Studies done on calcium seem to indicate dietary calcium is not as problematic as dicalcium phosphate, which can reduce manganese absorption (NRC, 2005, p. 238).

Iron and manganese compete at the same receptor sites in the body. Deficiency of iron increases absorption of manganese while supplementation of iron can decrease manganese in body tissues. This is particularly true of the inorganic iron (NRC, 2005, p. 238).

Deficiencies

Deficiency of manganese leads to impaired growth, skeletal abnormalities, ataxia of young and disturbed reproductive functions.

Sheep may exhibit difficulty standing, joint pain, poor locomotion and balance. In goats, tarsal joint excrescences, leg deformities and ataxia may occur (Underwood, 1981, p. 126). Excess calcium, iron or phosphorus may contribute to deficiencies of manganese, phosphorus, magnesium, iron, iodine, and zinc (NRC, 1985, pp. 13, 19; Underwood, 1981, p. 129). Low manganese may result in delayed estrus, low birth weights and abortions (NRC, 1985, p. 19).

Diets high in grains, soy or corn, for either poultry or ruminants, can lead to deficiency (another reason ruminants should not be grain-fed as a general rule) (Ammerman, Baker and Lewis, 1995, p. 239). Deficiency may actually increase copper absorption (NRC, 2007, p. 132).

There is some evidence that breed of poultry can make a difference in manganese needs (Ammerman, Baker and Lewis, 1995, p. 245). This is an interesting concept, as it seems to hold true for other breeds of livestock and other minerals as well. Because there can be variations in needs of different breeds, a sensible farmer considers what breeds can do best on his/her farm and management program, and continues to breed for animals that thrive under those conditions.

Manganese is one mineral where hair tests may be more reliable than anything else but most labs will not, at this time, do hair and wool testing for minerals and assume any accuracy (NRC, 1985, p. 19). One of the pitfalls of hair testing is the risk of environmental contamination of the hair or wool sample. Wool in particular will hold onto dust from the environment in the scales and must be processed carefully in the lab to remove contamination before testing. For this reason, most labs do not try to set parameters for mineral levels in wool. Bone is also used to test for manganese but this is much less feasible. Most testing, even for manganese, is done by liver tissue analysis. Most of the manganese in the body is stored in bones, liver or hair (NRC, 2001, p. 140).

Toxicities

Manganese is considered relatively safe at high levels, although pigs are more sensitive to toxicity than other species. Iron and manganese compete metabolically at the absorption site, excess manganese may contribute to iron deficiency (Underwood, 1981, p. 130). The symptoms most often noted in toxicity relate to iron deficiency. Manganese can interfere with other related trace minerals like zinc, copper and cobalt. Calcium and phosphorus also interact with manganese, toxicity of manganese can decrease utilization of calcium or phosphorus. Magnesium may also be impacted by excess manganese (NRC, 2005, pp. 238-239).

Manganese as a chelate is usually part of a complex that includes chelated zinc and copper and possibly some form of cobalt. This type of supplementation aims to decrease the negative interactions when one mineral is out of balance with another.

Pesticides sometimes contain manganese and some health practitioners believe this can cause a health risk. The risk may or may not relate to the higher manganese as much as to the toxic nature of the other chemicals in the pesticides. It is worth considering detoxification and support of elimination channels for minerals and for toxic chemicals if exposure to pesticides has occurred. This can happen anywhere but sometimes aerial spraying and lawn and home applications mean people (and animals) don't realize they've been exposed.

Ruminants

Deficiency can occur in ruminants on pasture but is more likely to occur in those animals that are grain-fed (NRC, 2007, p. 133). Symptoms of deficiency include improper bone growth, problems with fat and carbohydrate metabolism, reproductive issues and disrupted enzyme systems (specifically those that contain manganese). Superoxide dismutase is a manganese metalloenzyme and is an antioxidant (NRC, 2005, p. 235). Newborns can have ataxia (see also Copper) that is due to problems in development in the inner ear during gestation (NRC, 2001, p. 139). Bone abnormalities are caused by problems with enzyme reactions needing manganese (NRC, 2001, p. 139).

In studies, calves born to deficient cows had enlarged joints, bone deformity and weak bones, twisted legs and overall weakness. Deficient cows had abscesses on their livers and did not have proper amounts of bile in the gallbladder (NRC, 2001, p. 140).

Pigs

Manganese requirements in pigs are not well established. Deficiency can cause weak piglets, problems with bone growth, irregular estrus cycles (or none at all), fetus reabsorbtion and low milk production (NRC, 1998, p. 55).

Toxicity can reduce appetite and interfere with iron utilization in the body (NRC, 1998, p. 55).

Carnivores

There is no established requirement for manganese in dogs and a natural diet should supply all the manganese needed for dogs and other carnivores (NRC, 1985, p. 20).

Horses

So far, there is no established requirement for horses, although ruminant requirements are known. Deficiency in horses is assumed to be associated with structural changes in joints of foals (NRC, 1989, p. 17).

Iron (Fe)

Iron is a common, abundant mineral in the earth and vital to health. It is available as oxide, ferrous and ferric (ferrous being the best absorbed) as well as other chelated complexes. The ferr- component of the word relates to the old name for iron and is reflected in ferric, ferrous and even in ferrier (now farrier in American English), the tradesmen who make and place iron shoes on horses. The abbreviation for iron also comes from these old terms.

Oxidation reduction reactions in plants require iron and it is used during photosynthesis as part of carbohydrates (NRC, 2005, p. 199). In animals and people, iron is used in formation of hemoglobin (the basis of red blood cells), myoglobin (a protein in muscle tissue), enzymes and other biochemical reactions (Underwood, 1981, p. 69). Every cell in the body contains iron, it is found as transferrin in plasma, in milk (lactoferrin), liver (hemosiderin and ferritin) and in the placenta (uteroferrin). Adult animals of most species have similar concentrations of iron, although this differs in newborns, depending on species. Piglets are not born with much iron but adult pigs have approximately the same concentration as poultry, cattle, rabbits and dogs (NRC, 2005, p. 199).

In addition to the role of delivering oxygen to tissues, iron also acts as a cofactor in enzyme reactions important for cellular energy (Krebs cycle), electron transfer with cytochromes and carbon dioxide and oxygen exchange in hemoglobin (NRC, 2005, p.200).

In humans, iron-deficient anemia is common and an important consideration, but in animals, this is much less a concern. Some animals can be more susceptible but this is usually a result of unnatural conditions, such as veal calves raised in extreme confinement, pigs supplemented with high copper or newborn piglets from deficient sows. Otherwise, parasites that suck blood can cause iron deficiency (NRC, 2005, p. 200). This is most common in sheep and goats, less so in cattle.

Iron is absorbed in the small intestine but the rate of absorption depends on several factors. A healthy digestive system is needed for iron (and any mineral) absorption. Age and health of the animal impact how well iron is absorbed. Other supplements and nutrients interact with iron, making it more, or less, available. Milk proteins, phytate and soy protein decreases iron absorption. Lastly, what form the iron is in plays a large part in how well it is absorbed. Soluble iron proteinate forms are very available (NRC, 2005, pp. 200-201).

Like manganese, iron is bonded to transferrin for movement in the body. Remember that iron and manganese compete at these same sites. Most iron goes to bone marrow to be part of hemoglobin production. The remaining iron is used in enzymes (NRC, 2005, p. 200).

Excess iron is stored in the liver, spleen and bone marrow as ferritin and hemosiderin. When the body needs to use stored iron, a copper-containing enzyme is required. This is another way in which copper is needed for iron usage in the body. Iron is not removed readily from the body and what iron is removed in bile can be recycled back into the body. Some iron is also eliminated through urine, sweat, hair and nails (NRC, 2005, p. 201).

Plants contain varying amounts of iron, depending on soils, weather and plant species. Meats are high in iron and animal products (like oyster shells) can also be high in iron (NRC, 2005, p. 201). Some herbs can be good sources of iron, like chickweed (*Stellaria media*) and stinging nettle, and a wide variety of plant species in a pasture helps ensure that animals have access to as many trace minerals as possible.

Iron is absorbed from the diet in direct proportion to body needs. Because of this, animals with plenty of iron tend not to absorb iron well. Vitamin C (ascorbic acid) in the diet or digestive system helps the animal convert ferric iron, the less available form, to ferrous iron. This is a chelation process that occurs naturally (NRC, 2005, p. 202).

The form of iron most available for absorption depends on species. Carnivores make good use of iron from animal sources, like liver, while iron in grains is more available to birds but not well-absorbed at all in carnivores (Ammerman, Baker and Lewis, 1995, p. 172).

Deficiencies

Iron-deficient anemia occurs with, and can be hastened by, infections with parasites that suck blood, like *Haemonchus contortus*. Not all anemia is from low iron, pernicious anemia occurs when cobalt is deficient. Other minerals are needed for the body to use iron to make red blood cells, copper and cobalt in particular. See the section on Cobalt above for more information on pernicious anemia. Iron without copper is not useable. Iron and cobalt are used to make vitamin B12 in the digestive system, so in deficiencies, a B12 injection (or better yet, a B complex injection that also contains B12) is useful rather than iron supplementation.

Deficiency is associated with poor growth, lethargy, pale mucus membranes, increased respiration, decreased resistance to infection and death (Underwood, 1981, p. 70).

Polyphenols (like tannins), phytate and cadmium, copper, cobalt, manganese, nickel, calcium, phosphorus and zinc all interact with iron, making it less available (NRC, 1989, p. 17; NRC, 2005, p. 202). Phytate, however, can be a mixed blessing. Studies have shown that, in rats, phytate added to the diet increased iron absorption while in humans and calves, phytate-bonded iron was not available for absorption (Ammerman, Baker and Lewis, 1995, p. 173). Soy protein has been shown to decrease iron absorption (Ammerman, Baker and Lewis, 1995, p. 173). In my opinion, soy is not a great food source for animals or people.

Toxicities

Iron toxicity affects many organs and body systems, with extreme damage to the heart, pancreas and liver (NRC, 2005, p. 201). Too much iron in the diet contributes to reduced feed intake, reduced growth rate, inefficient feed conversion, anorexia, oliguria (decreased urine output), diarrhea, hypothermia, shock, metabolic acidosis, heart problems and death (NRC, 1985, p. 16; NRC, 2005, p. 203). Much of this damage may be due to potentially irreversible damage to the mitochondria, the powerhouses of the cells. Even mitochondrial DNA can be affected (NRC, 2005, p. 201).

Toxicity is associated with foal death and this presumably occurs in other species as well. Other toxicity symptoms include diarrhea, jaundice, dehydration, coma and death at any life stage. Destruction of the small intestine villi (important in nutrient absorption), lung hemorrhage and liver breakdown (from excess storage of iron in the liver) can occur (NRC, 1989, p. 17).

Potassium and vitamin E deficiencies can also occur with high iron. Excess iron may not increase body stores but can interfere with zinc levels in the body (NRC, 1989, p. 17).

Animals with enough antioxidants are less susceptible to iron toxicity. Older animals become more susceptible (NRC, 2005, p. 202). Large doses of iron can lead to death, although higher amounts orally over time are less of a problem because absorption is limited. Injection of iron should be avoided in animals unless recommended by a veterinarian, and dosage should be followed carefully. In many cases, low iron can be related more to low copper rather than actual iron deficiency. In these cases, injecting iron causes more harm than good and risks further depleting trace minerals and damaging tissues.

Ruminants

Deficiency causes weakness, lethargy, poor appetite, anemia, increased breathing rate and death. There is a difference in deficiency that is sudden rather than from deficiency of copper and cobalt interfering with iron absorption and usage in the body. Hypothyroidism can be associated with chronic iron deficiency (NRC, 2007, pp. 131-132).

In ruminants, deficiency is often a result of blood-sucking parasites and the body's inability to compensate for the lost blood, leading to sudden anemia and death. When this happens, there has to be a multi-step process to correct the problem. If the animal is to be saved, the parasites (usually *Haemonchus contortus*) must be stopped. Herbs or copper are a good choice here. See cautions above about sheep and copper. It is always worth re-evaluating pasture management and immune health of animals extremely affected. Second, a rebuilding must occur for the animal to make red blood cells. More than just iron is needed for this, adequate copper, cobalt and B vitamins are necessary for the body to use the iron to make hemoglobin. A diet with enough protein and energy is important (this does not mean grain!)

In *Haemonchus contortus* (barberpole worm) infection, a sheep can be dead within a week of parasite hatch from blood loss to these parasites. It is extremely important that shepherds keep a very close watch on sheep, breed for parasite resistance when possible, manage their pastures and grazing properly and provide herbs for deworming as necessary. Parasite prevention and treatment is covered in Volume 2.

Pigs

Because piglets can be born severely iron-deficient if the pregnant sow has low iron reserves, piglets are often treated with oral or injectable iron at birth. This can lead to toxicity and death in the piglets (NRC, 2005, p. 203). Rather than address this after the fact, it is safer and healthier to make sure sows receive proper nutrition during all life stages and particularly during pregnancy and lactation.

Carnivores

Although uncommon, toxicity can occur in dogs or cats if they have accidentally ingested mineral supplements intended for people. These accidental poisonings can prove fatal, so keep all supplements out of reach of pets (NRC, 2005, p. 203).

Horses

Horses, unlike ruminants, do not absorb iron well (NRC, 1989, p. 16). Toxicity can occur, however, if foals are given iron at birth. Foals can absorb iron well right after birth and have good transport systems in place, unlike pigs, so any supplementation risks toxicity (NRC, 2005, p. 203).

Several pasture plant species are high in iron and other trace minerals. This is a perfect example of why diversity in a pasture and rotational grazing in different areas can provide much benefit to herbivores. Trees like oak (*Quercus* spp.), hickory (*Carya* spp.), red cedar (*Juniperus virginiana*) and other trees and shrubs are not only high in tannins (reducing parasite loads) and allow for browsing off the ground, but they contain trace minerals brought up from the subsoil. Browsing on trees and brush keep animals eating higher off the ground, therefore avoiding ingestion of more parasites. Parasites do not climb well and generally are only an issue about six inches up leaves or blades of grass. Fruiting shrubs like bilberry (*Vaccinium myrtillus*) and lingonberry (*Vaccinium vitis-idaea*), blackberry (*Rubus* spp.), raspberry (*Rubus* spp.), hawthorn (*Crataegus* spp.), strawberry (*Fragaria* spp.) and rose (*Rosa* spp.) species are also astringent and contain iron. Weedy species like burdock (*Arctium* spp.), chicory (*Cichorium intybus*), comfrey (*Symphytum officinale*), dandelion (*Taraxacum officinale*), dock (*Rumex crispus* and *R. acetosella*), gentian (*Gentiana* spp.), nettle, parsley (*Petroselinum crispum*) and wormwood (*Artemisia absinthium*) contain many minerals and are useful for many reasons, including medicinal value (Levy, 1991, p. 171; Walters, 2013, pp. 229-230).

Iodine (I)

Our modern society is stressful, both physically and emotionally taxing, with many people lacking good coping strategies for stress. The vitamins and minerals needed to help the body deal with stress become even more important and often times, less available now. One of these "stress minerals" is iodine, needed for its role as part of the thyroid hormones. As for any other mineral, needs for iodine can differ depending on species, area of the country, life stage and amount of stress the animal (or person) is exposed to.

Iodine is available as iodates, the most readily absorbed form, and iodides (Pugh, 2002, p. 26). Supplemental forms are usually well absorbed, although ruminants do not make good use of diiodosalicylic acid (NRC, 2005, p. 186). Common in the ocean, seaweed and ocean life contain higher concentrations of iodine than are found in middle areas of continents. In some areas of the world, iodine in the soil can be so deficient that without supplementation, goiter develops (in humans and animals). Understanding the importance of iodine to the thyroid gland has contributed to the iodization of common products (like salt) to ensure everyone gets enough iodine for good health. Unfortunately, some of the iodized salts contain a synthetic form of iodine. Many drugs and chemicals, such as fertilizers, herbicides and pesticides, contain iodine. Sanitizers, such as teat and umbilical cord dips, often have iodine as well. These sources and supplements and other products can supply iodine to the body accidentally (NRC, 2005, pp. 182, 186).

Soils can be deficient in iodine and plant uptake of available iodine differs drastically by species, and even strain of plant. For example, cyanogenetic and non-cyanogenetic strains of white clover can differ by as much as 1200 ug/kg down to 40 ug/kg of iodine (Underwood, 1981, p. 86). Animal protein can be decently high in iodine, grains can be variable in their iodine content (NRC, 2005, p. 186). Where you are determines how much iodine is available in soils. Northern states in the U.S. are particularly deficient. Excess rain leaches iodine (and other trace minerals) from soils and glaciation removed iodine (NRC, 2005, p. 186). Some areas where nitrate concentrations are high in soils, however, have enough iodine: salt peter (potassium nitrate) deposits in Chile are one example. Contamination from industry or combustion exhaust from gas- and oil-powered engines can increase iodine in soils to high levels (NRC, 2005, p. 185). Evaporation of seawater also puts iodine into the air (NRC, 2005, p. 186).

Iodine is used in synthesis of thyroid hormones T3 (thyroxine) and T4 (triiodothyronine) that are needed to regulate metabolism (Underwood, 1981, p. 15). Remember that selenium is needed as well, for the body to convert T4 to T3. Tyrosine is also needed to form T4 that is then converted to T3 (NRC, 2007, p. 130). Without iodine and thyroid hormones, there can be no growth or cell activity (NRC, 2005, p. 182). The thyroid interacts in a feedback loop with many other gland systems (like adrenals and even reproductive glands). The hypothalamus, pituitary and adrenal glands are involved in a feedback with the thyroid to control almost all body functions. This is called the HPA-axis, and imbalance in this gland system is cited as the root cause of many of our modern chronic diseases. Low thyroid will mean an imbalance in other systems as well. This gland feedback system can be a problem in all animals and humans when it becomes out of balance. Stress, toxins and chronic illness will cause imbalances in the HPA axis and contribute to poor immunity and all types of chronic maladies.

It is not really possible to separate the function of the thyroid from the function of these other glands, although some practitioners try to do this, and focus only on one body system. This is a mistake, as it can cause things to become further out of balance. Dysfunction in this feedback system is the number one concern I see in practice. Although the disorders associated with it differ by person, the underlying imbalance is still the same and must be addressed before true healing can begin.

In animals, some of the same dysfunction occurs. An example of this is the autoimmune disorder Symmetrical Lupoid Onychodystrophy that occurs in some breeds of dogs. This is a result of an underlying imbalance in the HPA axis leading to autoimmunity.

Selenium, as was mentioned above, is important in thyroid function with selenium-containing enzymes being responsible for converting T4 to T3 (the active form of thyroid hormone). Thiouracil (found in the Brassica plant family) inhibits one of these enzymes, type I deiodinase. This enzyme is most important in newborn ruminants, less so in non-ruminant herbivores (NRC, 2005, p. 184).

Iodine is a supporting mineral for glands. It increases metabolism and helps reduce fatty tissue, helps safeguard the brain from toxins, acts as an antioxidant by removing toxic elements and promotes strong hair (Levy, 1991, p. 171). Reproduction, production of blood cells, regulation of body temperature, circulation and proper neuromuscular function all require iodine (NRC, 2005, p. 183).

Other nutrients that interact with iodine include vitamin A (to produce growth hormones) and manganese. Interference with iodine occurs when fluorine, excess calcium, cobalt, potassium or arsenic are part of the diet. Synthetic flavonoids can cross the placenta and cause problems with iodine and T4 production (NRC, 2005, p. 185).

Absorption takes place in the rumen in ruminants and both the inorganic and iodate forms of iodine are absorbed equally well. Storage is in the kidneys and thyroid gland, with iodine being found in almost all tissues of the body. Other areas where iodine is in higher concentration include placenta and ovaries, skin and hair, mammary and salivary glands (NRC, 2005, p. 184). The mammary glands and ovaries contain an active form of iodine while other areas of the body contain the inactive iodide (NRC, 2005, p. 184). Eighty percent of the iodine in the body is in the thyroid gland, the rest is stored in kidneys, with some in the heart, liver and muscle tissue (NRC, 2005, p. 189).

The body is able to store iodide and in some species (like humans), this stored source can supply the body for months. The body recycles iodine (as iodide) from breakdown of thyroid hormones and eliminates most of the excess iodine through the kidneys (NRC, 2005, p. 184).

Deficiencies

Iodine deficiency leads to goiter (enlarged thyroid). This enlargement is the thyroid gland's way to compensate for low iodine levels. A goitrous thyroid can somewhat compensate for the low levels, for a short time (NRC, 2005, p. 183). Other deficiency symptoms include weak or dead newborns at birth (see also Selenium and Copper deficiencies) and lack of wool on lambs. Adult sheep have reduced wool production and poor conception and animals can have dandruff and scurf (Underwood, 1981, p. 16). There will be depressed milk yield, pregnancy toxemia, abortion, retained placentas (see also Selenium), irregular estrus and infertility and depressed libido. Other mineral imbalances are also associated with the above symptoms.

The female fetus has a greater need for iodine and will die in utero, leaving an excess of male offspring (NRC, 1989, p. 16; Coleby, 2010, pp. 28-29).

Note that enlarged thyroid in goat kids may be congenital and not iodine-related (Pugh, 2002, p. 26).

In chickens, low iodine in the diet can lead to goiter without the reduced growth seen in other animal species, but eggs may have reduced hatchability (NRC, 2005, p. 183).

Availability of iodine can be depressed by methylthiouracil, a drug used to treat hyperthyroidism. Nitrates, perchlorates, soybeans and thiocyanates (potassium, sodium, molybdenum and mercury salts) also interfere with availability. These salts can be naturally occurring but are often a result of coal smelting. Other minerals that can contribute to availability are rubidium, arsenic, fluorine, calcium and potassium (Pugh, 2002, p. 26).

Legumes and brassica family are goitrogenic, unless heated. They will inhibit iodine usability, but increasing dietary iodine can compensate for this. There have been recorded instances of deficiency in newborns drinking milk from mothers who ate goitrogenic feed (NRC, 2005,

p. 183). As was noted above, other minerals in excess (like arsenic, fluorine, calcium, cobalt and potassium) interfere with iodine and thyroid hormones (NRC, 2005, p. 185). Other plants, like soybeans, also interfere with iodine. The cure for this is to feed extra iodine or remove the goitrogenic feeds from the diet. Studies show that adequate copper also helps the body better utilize iodine and offsets the negative effects of plants that inhibit iodine or thyroid hormone production (NRC, 2005, p. 188).

Requirements differ by soil content, climate and species. Seasonal differences in thyroid hormone production mean animals on summer pasture have decreased need for iodine (NRC, 2005, p. 183). This can make it difficult to get animals on pasture to come back and eat enough mineral to meet their needs. Farmers should make an effort to ensure their animals are eating enough mineral in summer.

Toxicities

Toxicity can also cause problems and species differences in tolerance to iodine are common. Iodine, however, is generally considered to have a wide margin of safety, as much as 1000 times the minimum in chickens and pigs (NRC, 2005, pp. 186-187). Toxicity can occur from supplemental forms like EDDI (ethylenediamine dihydroiodide) or kelp. Kelp is an absolutely fantastic source of iodine and trace minerals, but like anything, too much can lead to problems, just as can too little (NRC, 1989, p. 16; Coleby, 2010, pp. 28-29). Other products also contain iodine and can contribute to toxicity: disinfectants (umbilical cord treatment and teat dips), iodine tablets for water purification, some drugs and pharmaceutical products (NRC, 2005, p. 188).

In cattle, toxicity can occur from using EDDI, but generally cattle tolerate higher levels of iodine. The problem comes in because iodine passes into the milk. In dairy cattle, toxicity is less of a concern in the cows than it is for people using the milk, since people tend to be more sensitive to excess iodine (NRC, 2005, p. 187).

When people who have developed deficiency-induced goiter are supplemented with iodine, toxicity can occur because the thyroid has adapted to the lower levels, producing nodules that do not respond to excess by decreasing absorption. Toxicity is the result (NRC, 2005, p. 185). In some coastal areas of the world, like Japan, excess iodine consumption can cause goiter (NRC, 2005, p. 188).

Depression, anorexia, hypothermia, alopecia, goiter and poor body weight gain can be symptoms of iodine toxicity (NRC, 1985, p. 16). Note that goiter is a symptom of both deficiency and toxicity! Excess iodine can cause both hypothyroid or hyperthyroid function (NRC, 2005, p. 185). Too much iodine will decrease the body's absorption of iodine and increase secretion. In some cases, the problem may occur due to inorganic iodine excess but the form of iodine in excess may not be an issue in most cases. Iodide in excess can be one cause of lymphocytic thyroiditis or autoimmune thyroiditis (NRC, 2005, p. 185).

Selenium deficiency coinciding with high iodine can cause iodine toxicity damage in the body, while adequate selenium will offset the excess iodine and protect the body from deleterious effects (NRC, 2005, p. 188).

Plant sources include several tree and shrub species shoots, including red cedar (*Juniperus virginiana*), sweet gum tree (*Liquidambar stryaciflua*), oaks (*Quercus* spp.), hickories (*Carya* spp.), smooth sumac (*Rhus glabra*) and black walnut (*Juglans nigra*). Other sources include cleavers (*Gallium* spp.), garlic (*Allium sativa*) and seaweeds (including kelp) (Levy, 1991, pp. 171; Walters, 2013, p. 203).

Ruminants

Toxicity in cattle causes abortion, cough, nervousness, discharge from the eyes and nose, heart arrhythmias, skin problems, lowered milk production, hair loss and anorexia (NRC, 2005, p. 187). Sheep and other ruminants tolerate higher levels of iodine, for the most part, but excesses can lead to anorexia, cough, respiratory conditions, hyperthermia (overheating) and in extreme cases, death (NRC, 2005, p. 187). Iodine concentrations depend on iodine in the diet. Feeding of plants, like soybeans, can interfere with iodine availability. In the case of soy, this is due to interference with iodine recycling mechanisms, in the case of Brassicas, there is interference with uptake of iodine or T4 to T3 conversion. Plants that can be problematic include the aforementioned soy, rapeseed (*Brassica napus*), kale, cabbage and cruciferous (Brassica family) plants (NRC, 2007, pp. 130-131). Beet pulp, corn, sweet potatoes, millet and white clover can all interfere with iodine (NRC, 2001, p. 137).

Beet pulp has become a stable in some circles for feeding to ruminants not on pasture or hay. While this may be a short-term solution to isolated situations (like severe drought limiting pasture and hay), it should not be a long term feeding strategy, in my opinion. With the advent of GMO sugar beets, it will become increasing difficult to find non-GMO beet pulp for feeding. Ruminants were designed to forage and farms should be set up so they have access to pasture with as much diversity as possible during growing seasons.

Sheep and goats require similar levels of iodine and all ruminants tolerate high levels of iodine (NRC, 2007, p. 130).

Deficiency can occur when soils are deficient and there is not adequate supplementation, or when plants that interfere with iodine in the body comprise a large portion of the diet. Other problems that can interfere with iodine or cause deficiency include stress. This includes cold stress in goats, which can be a result of deficiency as well. Iodine is important for reproduction. Deficiency in cattle, sheep and goats can lead to decreased lambing and kidding percentages in sheep or goats, stillborns or hairless newborns and infertility in all species (NRC, 2007, p. 130; NRC, 2001, p. 139).

Rabbits

Rabbits fed a natural diet can have kelp as their source of trace minerals and salt. Depending on the breed and farm situation, the rabbits may require a different mineral source as well but kelp can be a good way to supply needed salt and has the advantage of providing other minerals in a natural, bioavailable form. Blocks that contain high salt, molasses and additives should be avoided but a small bowl of kelp is inexpensive and easy to provide.

Pigs

As was mentioned above, pigs tolerate iodine well and excesses are not usually an issue. At very high levels, pigs had lowered growth rate and hemoglobin production was reduced. This may be due to iodine interfering with iron stores in the body (NRC, 2005, p. 187).

Carnivores

In cats, excess iodine can cause problems with thyroid hormone, leading to hyperthyroidism and possibly carcinoma of the thyroid (although studies are not clear on this) (NRC, 2005, p. 188). Commercial pet foods can be high in iodine (and salt), high enough to cause toxicity issues in carnivores (NRC, 2005, p. 188). Homemade diets, on the other hand, may benefit from the addition of a small amount of kelp as a natural iodine and trace mineral source, unless seafood is a large part of the diet.

Horses

Horses have less tolerance for toxicity than ruminants and can have typical symptoms of toxicity (NRC, 2007, p. 130). Toxicity in horses can also lead to osteoporosis, problems with phosphate levels, weakness in the legs and goiter in foals (NRC, 2005, p. 187).

Deficient foals usually have goiter, are weak and have trouble nursing. Mares have problems with fertility (NRC, 1989, p. 16). Horses, like other animals, enjoy kelp and this can be a good source for iodine.

Zinc (Zn)

Zinc is used to galvanize iron and water tanks for livestock. It is worth noting, however, that if the water is acidic, zinc can leach from the tank into the water supply (NRC, 2005, p. 415). Zinc is also used in ointments and overused in supplements for men. Taking zinc supplements or lozenges (to prevent or treat colds) can eventually cause imbalances in the body. A note about men and prostate health: zinc is important for prostate health, but supplementation of zinc can throw off levels of related minerals, like copper. Zinc should not be supplemented as a single mineral. In cases where men have used only zinc as a supplement for prostate, sideroblastic anemia can occur (NRC, 2005, p. 418). In sideroblastic anemia, red blood cells that are misshapen when produced in bone marrow cannot use iron to form hemoglobin for oxygen transport. If zinc must be used as a supplement, a multivitamin that also contains copper in the proper ratio to zinc is a safer option or, increase zinc-containing foods in the diet. Fifteen parts zinc to one part copper is one recommended ratio for health.

Pollution can be a significant source of zinc in soils, including zinc in fungicides and pesticides (NRC, 2005, p. 415). Pollution from mining and industrial dumping can contaminate water with high levels of zinc (NRC, 2005, p. 415). This can also be an issue for other minerals as well, including copper and phosphorus. As was noted above, galvanized containers can leach zinc into water or food, especially under acidic conditions.

Zinc is usually higher in legumes (lower in cereal grains) but the high calcium in legumes may offset the higher levels of zinc available (Pugh, 2002, p. 26). Legumes also contain phytate, as do seeds and grains. Phytate binds to zinc and causes its complete elimination (NRC,

2005, p. 416). While it is commonly assumed that fiber interferes with zinc absorption, in fact it is phytate that is problematic, although many plants contain both (NRC, 2005, p. 416). As plants mature, they do not uptake zinc as well from soils. Fertilizers (like lime or superphosphate) decrease availability of zinc in soils. These fertilizers may cause other imbalances, besides not being a particularly natural way to restore the ecosystem and soil balance. Animal products, like fish or meat meal, tend to be higher in zinc while grains are lower. Soy, peanuts and linseed fall in the middle (NRC, 2005, p. 415). Zinc in seeds and grains may not be well-absorbed however, making these sources somewhat questionable (Ammerman, Baker and Lewis, 1995, p. 373). The calcium content of the diet also relates to availability of zinc and is often considered in conjunction with phytate (Ammerman, Baker and Lewis, 1995, p. 373).

The form of zinc used in supplements can vary widely, zinc oxide is often used but may be subject to interference from antagonistic or related minerals that are not in balance. Zinc is also available as zinc citrate, proteinates and amino acid complexes (the chelated forms), and sometimes as zinc carbonate, sulfate or chloride (NRC, 2005, p. 415). Some studies in pigs show that chelated forms of zinc are better absorbed than other forms, but other studies have shown mixed results. In studies with chicks, the chelated forms were much better absorbed (over 100% higher than zinc sulfate) (NRC, 2005, p. 416). Much of the confusion may arise from the complex interactions of zinc, other minerals and food in the digestive system. Minerals are not isolated in the body and their use and absorption require other minerals and vitamins. This continuous interaction makes isolating minerals and results in real life more difficult. In vitro studies do not often compare well to in vivo studies, either.

Absorption takes place in the rumen and small intestine of ruminants, but only the small intestine in non-ruminants. How much zinc is available for absorption depends on several factors. Excess iron, copper and calcium or phytate decrease absorption, and proper amino acids enhance absorption. Excess iron, even iron supplementation in the diet, has been shown to decrease zinc absorption; however, slightly excessive copper does not seem to be as problematic. Cadmium and tin can also interfere with zinc absorption (NRC, 2005, p. 415). Amounts absorbed from the diet can vary from fifteen to sixty percent, with young animals absorbing zinc more efficiently than older animals (NRC, 2005, pp. 414-415). Zinc is excreted through the feces, including amounts released from the body and zinc that was not absorbed to begin with. During times of stress and trauma, zinc is eliminated in urine but this amount is usually very low (NRC, 2005, p. 414).

Many enzymes in the body contain zinc, and DNA replication requires zinc, as does growth in the body (NRC, 2005, p. 413). It is important in metalloenzymes in blood and tissues, examples include alkaline phosphatase, liver alcohol dehydrogenase, pancreatic carboxypeptidase, thymidine kinase. It is found in erythrocytes, leucocytes, platelets and blood plasma (Underwood, 1981, pp. 138-139). Concentrations occur in the choroid and iris of the eye and in the prostate gland. There are lower levels found in skin, liver, bones and muscle tissue, blood, milk, lungs and brain (NRC, 1989, p. 18).

Protein, carbohydrate and essential fatty acid metabolism all depend on zinc, as does vitamin A synthesis. Deficiency can lead to anorexia (see also Cobalt and Sulfur) (NRC, 2007, p. 137). The immune system depends on zinc, one of the antioxidant minerals. Chelated zinc has been shown to improve milk quality and udder resistance to stress (NRC, 2007, p. 137).

Important for a healthy reproductive system (both male and female) and prostate gland health, there is no difference between amounts needed for male and female (Coleby, 2006, p. 113; NRC, 1989, p. 18; Coleby, 2010, p. 37).

Need for zinc increases during stress, illness and fasting. Levels of zinc in plasma can be under hormonal control and involve amounts stored and released from the liver (NRC, 2005, p. 414). During times of disease or immune system stimulation, zinc is moved into the liver, thymus and bone marrow rather than the skin, intestine or bone (NRC, 2005, p. 414).

Deficiencies

Deficiency can produce dermatitis and parakeratosis, particularly on legs in horses and legs, feet and faces of llamas. Alopecia can occur as well. Alkaline phosphatase (ALP) is an enzyme important in bone health, in addition to other functions in the body. During times of deficiency, ALP activity is reduced. Zinc levels in serum and tissue samples are also decreased. There can also be decreased milk production, impaired appetite, poor food utilization and subsequent slowed growth during deficiency. Increased susceptibility to foot rot, less hair on legs and head, swollen joints and decreased reproduction (including defective spermatogenesis) are all associated with zinc deficiency. Impaired vitamin A metabolism and reduced testicular development occur with deficiency. Excess salivation, wool loss and delayed wound healing can occur.

Male goats, and camelids, are more susceptible to insufficient zinc (Pugh, 2002, p. 26; Coleby, 2006, p. 113). This can manifest in llamas and alpacas as a loss of hair on the bridge of the nose and the toes. Some people mistakenly believe this is due to sunburn but the underlying cause may relate to zinc.

Marginal deficiency may be more common than larger deficiency. Studies have shown reproductive and weight gain from addition of zinc can be achieved in deficient areas. Increases in number of lambs are seen when animals are supplemented with zinc (Underwood, 1981, p. 136).

Calcium, phosphorus and phosphate fertilizers can reduce zinc availability. Excess zinc interferes with copper but the reverse is as problematic (Pugh, 2002, p. 214).

Since zinc is related to vitamin A metabolism, zinc deficiency also leads to vitamin A deficiency (NRC, 2007, p. 138).

Toxicities

True toxicity is rare, possibly one gm/kg feed can cause reduced feed intake, reduced weight gain and lead to hemolytic anemia (Jain, 1993, p. 200; NRC, 1985, p. 20; Pugh, 2002, p. 26). Most animals tolerate high levels of zinc fairly well. For the most part, zinc is considered to be non-toxic (NRC, 2005, p. 416). Of course, this can vary depending on circumstances, species and amounts of zinc. In no way should zinc be used excessively just because science has yet to find much damage from overuse. In cases where toxicity is known to occur, the mechanisms are not well understood yet but involve DNA, enzyme and protein systems in the body. Excess zinc causes deficiency in copper and the immediate result of excess is irritation to mucus membranes in the digestive system. Toxicity in ruminants can lead to die-off of rumen bacteria in addition to deficiency of related minerals (NRC, 2005, p. 415).

Zinc in the environment can lead to toxicity, contaminated water, feed or air pollution being the most common sources of contamination. Symptoms of toxicity include anorexia and generally unthrifty animals (NRC, 2005, p. 415). In cases of sudden high levels of zinc in the diet, symptoms of toxicity can be vomiting, diarrhea and cramps (NRC, 2005, p. 416). In dogs, hemolytic anemia can occur followed by organ failure (NRC, 2005, p. 416). Sources of large amounts of zinc include batteries, galvanized metals (which can include cage wires and nuts used to hold cages together) and pennies which were minted after 1982 when copper was lessened and zinc increased (NRC, 2005, p. 416).

Boluses in livestock can be a source of possible toxicity and should be used judiciously, if at all (NRC, 2005, p. 416). There are a few occasions where boluses may be an evil necessity, but in general, they are not a natural way to supplement minerals and can have downsides.

This is mentioned throughout the text, consider all the angles to bolus use before integrating them into a natural management system (if you can call boluses natural).

Copper deficiency related to excess zinc can lead to copper-deficient anemia. It has been noted several times above that not all anemia relates to deficiency of iron, and even when it does, there can be underlying mineral imbalances keeping iron from being available. In toxicity, immunity is reduced and the pancreas may be damaged or fail. The pancreas appears to be more sensitive to zinc toxicity than other organs but the mechanisms for this are not yet fully understood. Accidental zinc overdose in calves led to a range of symptoms, including problems in the eyes, pneumonia, diarrhea, anorexia, heart arrhythmias, bloat, convulsions and death. These symptoms occurred in calves that were still milk-fed (NRC, 2005, p. 416).

A note here about testing for mineral levels: blood stored in purple topped vials can be contaminated with the zinc that is used in the purple stoppers. Any blood drawn for mineral analysis should be stored in royal blue vials. If your veterinarian mistakenly grabs the wrong vial, please remind them of this and ask for the blue top vial instead.

Ruminants

While ruminants can often be lumped together when discussing mineral needs or mechanisms of absorption or storage in the body, in the cases of zinc and copper, sheep are different. Because sheep are unique in how they use, store and release copper, the problems with zinc toxicity change a bit. Sheep fed excess zinc have copper deficiency but unlike other animals, feeding higher levels of copper does not restore the copper levels in the body. Zinc toxicity symptoms (problems with weight gain and feed use and lamb survival) persist even with adequate copper to compensate. In other animals, like chickens, when excess zinc interferes with copper in the body, injections of copper can restore the copper levels in tissues and storage (NRC, 2005, p. 417).

Zinc is absorbed in the small intestine based on need, and absorption is fairly high in ruminants compared to some trace minerals. As much as sixty to seventy percent of dietary zinc can be available in sheep and goats unless there are interfering minerals like selenium, cadmium, copper, calcium or metallothioneins (NRC, 2007, p. 137). Metallothionein, the same protein that transports copper, also transports zinc and cadmium. Excesses of either of those minerals will result in less transport of zinc, either for elimination or movement into the body from the liver. Excess zinc is eliminated in feces. Zinc is not stored well in the body but deficiency may be delayed due to movement of zinc from muscle and bone. Sheep wool can be high in zinc but is not necessarily an easy source for mineral testing due to the scales accumulating environmental toxins. Red blood cells also contain zinc (NRC, 2007, p. 137). Stored zinc in the liver is not particularly available to be brought back into circulation to prevent deficiency. Animals with normal liver levels of zinc may still exhibit deficiency symptoms during times of dietary deficiency (NRC, 2001, p. 146).

Toxicity may be more of an issue for sheep than for other animals, possibly relating to their inability to remove these types of minerals with metallothionein. Toxicity in sheep causes problems with rumen function, eating of wool (pica), lack of weight gain and anorexia (NRC, 2007, p. 138).

Adult females with adequate zinc pass this on to nursing young through the milk (NRC, 2001, p. 144).

As with other species, excess zinc interferes with copper and extremely high amounts of copper could potentially interfere with zinc absorption (although this may not be common in cattle). Iron also interferes with zinc, as does calcium and phytate. Cadmium or lead can interfere with zinc (NRC, 2001, p. 145).

Some studies seem to show that ingredients in feed may interfere with zinc but what exactly are the problem nutrients and how this occurs is not yet known (NRC, 2001, p. 145).

Pigs

For the most part, pigs tolerate higher levels of zinc (NRC, 2005, p. 417). Zinc deficiency (known since 1934) and the need for supplementation in farm animals was recognized in 1955 in part due to problems of parakeratosis in pigs fed diets high in calcium and phytates that interfered with absorption of dietary zinc (Ammerman, Baker and Lewis,1995, p. 367). Parakeratosis appears as thickened rough areas of skin that can itch and are susceptible to infection, causing even more problems.

When pigs have access to very high levels of zinc, much of the possible toxicity can be offset by adequate iron (NRC, 2005, p. 417). Diets higher in soy and corn can increase the requirement for zinc due to interference from phytates. Barrows (castrated males) have the lowest need for zinc, gilts (young females) need more than barrows and boars have the highest requirement, possibly due to the reproductive needs in adult males (NRC, 1998, p. 56).

Deficiency, as was noted above, causes parakeratosis but can also prolong birthing (farrowing) and interfere with sperm production, milk production and uterine health. In addition, low zinc levels interfere with the enzymes related to zinc, reduce the number of piglets born and cause their weight at birth to be lower. Boars will have problems with the testes and piglets will have problems with thymus development (NRC, 1998, p. 57).

Some of the elemental zinc forms are fairly well absorbed in pigs but zinc oxide is not as available for digestion. Chelated zinc is also bioavailable but zinc from plants or grains is not readily digested (NRC, 1998, p. 57).

Pigs tolerate high levels of zinc with no ill effects, although continuous feeding of these amounts will produce joint problems and general poor condition. Extremely high levels (up to 4000 ppm) leads to arthritis, depression, digestive system irritation and death. Other effects of high zinc levels include copper deficiency (NRC, 1998, p. 57).

Carnivores

In dogs, zinc has not been studied as much as other minerals. Excess calcium may interfere with zinc and lead to deficiency symptoms such as emaciation, skin conditions, hair loss, vomiting and poor condition overall. The liver is adversely affected and kidneys can have calcium crystal damage (NRC, 1985, p. 20).

In dogs with health problems, zinc deficiency can occur as well. Hypothyroidism, diseases, liver problems or invasive surgery, like spaying, is linked the zinc deficiency (NRC, 1985, p. 20). This should not be surprising since zinc is so closely linked to immunity and cellular function.

It is important to note that high-calcium diets are not abnormal for dogs or carnivores in general. Carnivores eat bones. Making sure your carnivores have access to meat and organs in addition to some real foods should provide the wide range of nutrients necessary for good health. Adding some kelp to the diet can help with zinc, kelp contains zinc and other beneficial minerals, including iodine.

Horses

Horses tolerate high amounts of zinc, but some symptoms can occur. Bog spavin (tibiotarsal swellings reported in Arabians with high zinc) and foals developing lameness, stiffness and copper deficiency can occur with excess amounts of zinc (NRC, 1989, p. 18). In some cases, this toxicity can occur from environmental pollution of zinc into ground where horses are pastured (NRC, 1989, p. 18).

Deficiency in foals leads to anorexia and slowed growth rate, hair loss, parakeratosis and decreased zinc-related enzymes (NRC, 1989, p. 18).

Plant sources

Zinc is one of the minerals found in high amounts in seaweeds, several fruit and nut trees and rose hips. Tree stems that contain zinc include black cherry (*Prunus serotina*), sweetgum tree (*Liquidambar styraciflua*), buckbrush (*Symphoricarpos orbiculatus*), persimmon (*Diospyros virginiana*), leaf and stem of sassafras (*Sassafras albidum*) and vegetables like cabbage and tomato (Coleby, 2006, p. 113; Walters, 2013, p. 280). There are other trees, shrubs and vegetables that contain decent amounts of zinc, a varied pasture is a great source of trace minerals.

Silicon (Si)

This mineral has not been well-studied and is not included in requirement recommendations for herbivores. Silica, however, is most certainly a necessary mineral for good health. It is used in collagen, skin, hair and joints. Silica promotes and maintains strength of hair, hardens teeth and nails, strengthens eyes, and contributes to the suppleness of limbs, ligaments and skin (Levy, 1991, p. 172).

Silica is found in varying amounts in many plants, particularly grasses, but herbalists tend to use herbs like horsetail (*Equisetum* spp.) in formulas to support and heal the above-listed body systems. *Equisetum* fresh, however, can have a thiamine-inhibiting (vitamin B1) action although the act of tincturing, heating for tea or drying for hay mitigates this effect.

Equisetum species are coarse, primitive plants that either grow whorls of stem-like leaves or a straight jointed stalk, or both, depending on if the plant is in fruiting stage or not.

Field horsetail is not generally considered fatal to sheep, although it may cause digestive upset, but can be fatal to horses. Poisoning occurs, supposedly, only if the horses are fed twenty percent of their daily feed as horsetail. That's a lot of horsetail! Removal of the horsetail results in reversal of the poisoning symptoms in horses (Bebbington, 2013).

Grieve's "A Modern Herbal" records that *Equisetum hyemale* was used as horse fodder in Sweden, although cattle avoided it, and it was thought to wear down their teeth. Reindeer will eat certain species and choose it over hay (Grieve, 1971, p. 420-421).

In Fyles' "Principal Poisonous Plants of Canada" (1919), a description of the symptoms of poisoning are given:

"The first general symptoms are a certain excitement, unthriftiness, diarrhea, good appetite; later, staggering gait, partial loss of motive power, craving for the weed, pulse accelerated, respiration difficult, sometimes convulsions and death or a state of unconsciousness and coma. Sometimes the attack is very acute, death occurring in a few hours; usually, however, the disease lasts from a few days to several weeks."

Fischer-Rizzi in "Medicine of the Earth" (1996) records an old saying that horsetail was considered "Horses' bread, cows death." Which, of course, contradicts some of the above sources. She records similar symptoms for staggering disease: excitability, cramps, disturbed coordination, death due to paralysis (p. 174).

Fischer-Rizzi also records that the staggering death associated with horsetail (in sheep, horses and cattle) can be attributed to equisetine, a chemical deriving from the parasitic fungus infecting horsetail. The fungus, *Ustilago equietti* is especially common on *Equisetum palustre, E. silvaticum, E. hyemale*. Fischer-Rizzi lists *E. palustre* as the one growing in swamps and bogs. This is the one associated with the staggering disease, according to that author (p. 173).

Brown spots appear on the stems and whorls where the fungus has infected the plant. In northern Minnesota, this occurs later in summer, early *Equisetums* are generally free of the fungus. Only harvest plants free of discoloration, insect damage or wilting and avoid harvesting in areas where pollution could have collected on or in plants. *Equisetum* can accumulate some other trace minerals, like selenium, if present in the soil.

Boron (B)

Boric acid is a common form of boron in animals and people. Boric acid or borax are common supplements. Legumes and many other plants need boron to grow. Boron is used in bones and brain health and immunity, among other functions. Boron is also needed for the body to use calcium, magnesium, molybdenum, selenium, omega 3 fatty acids, proteins and vitamins A and D (NRC, 2005, p. 61).

In spite of the fact that this mineral is known to be essential to plants and to have health benefits in animals, it is usually not studied or considered when assessing mineral imbalances. It can be deficient in the eastern U.S. in soils, particularly sandy or silt loams (NRC, 2005, p. 60).

Ingested boron is rapidly excreted in urine, after absorption from the digestive system. In a few instances, boron can be absorbed through skin if there are open wounds. This can lead to toxicity, but is more rare now that boron-containing topical products are not common. In cases of toxicity, vitamin B2 (riboflavin) can reverse the toxicity (NRC, 2005, p. 61).

Boron may provide benefit in toxicity from other minerals, including toxicity of fluoride (NRC, 2005, p. 61). Fluorine is not included in this section since it is generally not included in supplements. There can be issues related to toxicity of fluorine, however.

Use of boron in supplements is growing. Treatment for postmenopausal symptoms, arthritis and osteoporosis, in both people and animals, has been done in countries besides the U.S. and in growing levels in the U.S. (NRC, 2005, p. 62). Supplementation may, or may not, be needed for otherwise healthy animals and people since most boron in the diet and environment is available for absorption.

Toxicity can be a concern if using supplements. This used to occur accidentally when boron was used more commonly for food preservation (NRC, 2005, p. 64). Symptoms include ataxia, kidney damage, convulsions, lowered body temperature, violet-red skin and mucus membranes for various animal species (NRC, 2005, p. 61). In people, toxicity can produce digestive pains, vomiting and diarrhea with headache, lethargy and lightheadedness. Rash is a less common symptom. Kidney damage and death can occur but this is extremely rare and would almost certainly be a result of accidental overdose or poisoning at a high level (NRC, 2005, p. 63).

Mineral Interrelationships In Animals

Minerals interact with each other continuously in the body, excesses of one or more minerals can interfere with absorption of other minerals. Below are some of the more common interactions that potentially cause problems in the body:

Calcium in excess can interfere with magnesium, phosphorus, sulfur, manganese, zinc and iron.

Phosphorus in excess can interfere with calcium, magnesium, iron, aluminum, manganese, copper, molybdenum and zinc.

Magnesium in high amounts interferes with calcium, phosphorus, potassium and manganese.

Sulfur interacts with calcium, selenium, copper, molybdenum and zinc.

Sodium and potassium interact with each other, causing issues when either is out of balance.

Potassium interferes with magnesium absorption as well.

Zinc can interfere with calcium, phosphorus, sulfur, iron, cadmium and copper.

Iodine interacts with iron, cobalt and arsenic.

Molybdenum interactions with phosphorus, sulfur and copper; molybdenum binds with sulfur and copper to form a complex completely unavailable for digestion.

Manganese in excesses interacts with calcium, magnesium, phosphorus and iron (it binds to the same receptor sites in the body as iron).

Iron interacts with phosphorus, manganese, zinc, cobalt and copper.

Selenium in excess can cause issues with sulfur and arsenic.

Copper has interactions with phosphorus, sulfur, iron, cadmium, zinc, molybdenum and silver.

Source: http://www.aces.edu/pubs/docs/A/ANR-0890/ (Originally Adapted from Nutrient Requirements Of Beef Cattle, Sixth Edition, 1984)

Putting It Back Together

The above chapters on vitamins and minerals leave most of us with the assumption that each substance can be considered separately. This is not correct. These nutrients interact with each other in soils, in plants and in animals. Weather, climate and geography all affect what's available to animals. Seasonal differences of plants' needs can change what is available in pastures and hay. As animals grow and start to reproduce, their needs change and their bodies' ability to absorb nutrients change as well. For herbivores, supplemental mineral mixes are almost a necessity, but the labels and ingredients can be quite confusing. Even certified organic products often contain ingredients that serve no purpose other than to carry trace minerals or add bulk to the mix for ease in mixing and handling.

What I have found over the years is that helping people formulate their own mineral mix is often more work for them than it is for me. The tediousness of making your own mix (and the problems associated with trying to mix trace minerals in a five gallon bucket) almost guarantee the animals will not get their mineral supplements regularly. It's just too much work. I encourage most people to look carefully at labels, work on managing their farms better and make sure ruminants in particular have enough cobalt, copper and selenium (with exceptions for breed-specific concerns with copper). My own mineral line Back in Balance Minerals® is available in the U.S. for those interested: backinbalanceminerals.com. When choosing a mineral supplement for livestock, read labels, read labels and then read the labels again. Look for proper ratios of related minerals in forms most easily digested. Look also for additives and excess salt, molasses or odd ingredients like feathers, which really have no place in a supplement for a herbivore. Some supplements make use of the fact that ruminants can convert ammonia to protein in the body and urea blocks are available to increase protein. My suggestion, if you are looking to increase protein, is to look at pasture management and natural sources of protein, like legumes. Do not underestimate the protein content of browse and some grasses during parts of the year. These sources can provide as much protein as alfalfa.

"Let your food be your medicine and your medicine be your food" is just as applicable to animals as it is to people. The best way to ensure healthy animals is to supply them with healthy food sources appropriate for the breed. Use natural supplements to fill in any gaps in the feeding and pay attention to health. Ill health in all its forms is a clue to underlying imbalances. Once you learn to see disease and parasites as the symptoms rather than the problems, you can start to adjust your thinking and your feeding to promote health.

For information on natural ways to restore health to already diseased or parasite–ridden animals, see Volume 2. You can begin right now to work to restore natural balance in the digestive system according to the information given above. Feeding according to species differences using natural, organic foods in their whole forms will restore and maintain health in many situations without the need for further intervention. In conditions where there has been damage done to the body, natural therapies to restore tissues and organs while returning to a natural diet will drastically improve the health and well–being of animals (and people!).

For carnivores, mineral supplements in most cases are unnecessary. This assumes, of course, that you are feeding a real-food diet, not kibble. When animals are ill or elderly, supplementation can be useful but should be done with caution. It is always better to let the diet provide the needed nutrients. This requires some care in selecting organic or sustainably raised meats and foods for your carnivores. Elderly carnivores do have trouble switching to a raw food diet. Preparing lightly cooked or steamed foods may be necessary and raw bones might need to be ground or crushed in some way to spare teeth and jaw muscles that are no longer strong enough. In the wild, carnivores don't survive to extreme old age because they are unable to continue to hunt and eat their food, but as our pets, some simple measures in the kitchen will ensure your older carnivores can continue as your companions for many years.

CHAPTER 11

BIRTHING AND YOUNG ANIMALS FROM A NATURAL PERSPECTIVE

Lambing and kidding naturally is a very easy process. The farmer's role can be critical because the role involves making sure the mothers have had the best nutrition and access to a variety of plants during gestation. Once birth commences, the farmer's job becomes one of observation. Ewes and does (female goats) should give birth easily to twins or triplets (or even occasionally quads) unassisted. Once the young animals are born, the mother knows what to do. No interference is necessary and may even be detrimental to normal bonding and suckling. A frightened ewe or doe won't let her milk down and the additional hormones in response to stress will come through the milk. Young animals should be on their feet right at birth and looking for the teat. Even a ewe in long fleece should be able to suckle her young, the lambs find the milk by smell, not sight. Allowing the ewe or doe to stay with her flock and in her normal environment reduces stress and allows other older mothers to act as teachers if necessary. Usually, an animal going into labor will seek some solitude to give birth and won't reintegrate with the flock until her young are fed and ready to run. This can be a few hours but isn't long. With proper vitamins and minerals given to the mother during gestation, colostrum will provide all the nutrition and energy necessary for the young animal to be off to a healthy start and ready to become a strong, productive member of the flock.

It is generally not necessary to treat umbilical cords, mothers clean that off while cleaning birth fluids off the newborn. If you are unable to sleep nights without treating the cord in some way; however, I recommend an essential oil blend or powdered goldenseal (*Hydrastis canadensis*) as a safe, natural treatment. Goldenseal was never a common woodland wildflower and has been over-harvested in our forests. Please choose certified organic goldenseal to help conserve this lovely native medicinal. If using essential oils, please dilute the oils in a carrier. Water is fine but you need to use an emulsifier, like a few drops of vodka, to get the essential oils to mix with the water at all. In a four-ounce glass spray bottle, up to fifty drops total of essential oils can be used. More information on essential oil use will be given in the next volume. Sources for herbs, essential oils and supplements are given at the end of the book.

Proper Digestion/Beneficial Bacteria and Probiotics

Human infants and other young mammals are "inoculated" with beneficial bacteria from their mothers' colostrum, in some cases from regurgitated foods from parents and also from environmental exposure. When a young animal is raised without access to colostrum, or is given pasteurized colostrum, proper digestive function can be delayed or even inhibited and the young animal may die.

Can Early Vaccination Upset Beneficial Bacterial Balance in Young Animals?

Studies done on human infants and mice are showing that the newborn has lowered immunity to encourage establishment of beneficial microbes in the digestive tract. The immune system develops over time, allowing beneficial microbes to be fully established before full immunity is reached. Early use of immune system stimulation and other disruptions to this delicate balance risk inhibiting beneficial bacteria in the digestive tract (Elahi, et al, 2013). Domestic animals are now being vaccinated earlier and more often. Could this play a role in reducing beneficial bacteria in the gut and contributing to poor immune function? Hopefully further studies will help us understand better the benefits of the natural system and the risk of early disruption.

Antibody Transfer in the Newborn

Some species transfer antibodies (large protein molecules that react with antigenic substances) from the placenta, others require that the newborn take in colostrum. There are three classes of immunoglobulins: IgG, IgM, IgA. Adult intestines cannot absorb the larger protein molecules intact (except in some cases of disease) but newborn intestine can. Pigs, horses and ruminants receive all their maternal immunoglobulins from the colostrum. These young animals are described as hypogammaglobulinemic at birth. A properly functioning immune system in young of these species absolutely requires colostrum! Pasteurization destroys the immunoglobulins, colostrum should not be pasteurized. Rodents gain immunity both from placental transfer and colostrum. Absorption of the large protein molecules stops in the intestine in response to factors in the colostrum.

The Importance of Colostrum and Mother's Milk. Replacers When Necessary, What's Best?

It is unconscionable to remove a young animal from its mother if there is no other reason than just to have a pet, or allow children to bottle-feed a cute animal. In cases where there is a true emergency, real milk replacers should be used (see below). Both mothers and their young go through emotional and hormonal changes if they are separated before natural weaning. If it is necessary to separate them, consider using flower essences and calming essential oils to help mitigate the stress response to this unnatural event.

Commercial powdered milk replacers have several drawbacks and are not recommended as part of a natural, healthy start to raising orphaned young. The commercial replacers do not introduce beneficial organisms to the digestive tract and young ruminants fed these replacers are at risk for bloat at the age when their rumens start to function as adults (four weeks in lambs). This is in addition to the potential harm from chemicals, preservatives and genetically modified ingredients (GMOs) found in most milk replacers. These replacers also do not contain the needed vitamins and minerals in forms most available for digestion. Young animals raised on replacers are also at risk for deficiency diseases.

Some symptoms of deficiency include young lambs and kids crawling on their knuckles. Young can be thin with poor hair coats. Legs may bow or be crooked and the animal may be unable to stand properly or be too weak to keep its head up for nursing. Switching to a real milk replacer and supplemental minerals when needed to combat deficiencies will correct these problems. Sometimes additional selenium and vitamins E, A and D are necessary. Cod liver oil can provide the needed vitamins A and D. Twenty cc for lambs and kids is appropriate for a few days. Do read the label on the cod liver oil. It should be filtered for heavy metals and provide approximately 4000 IU vitamin A and 400 IU vitamin D. See under Vitamins and Minerals for more information on cod liver oil. Selenium and E can be given in a human gelcap form. The chelated selenium/vitamin E supplement can be used in a pinch. One to two gel caps with 400 IU vitamin E and seventy micrograms (mcg) selenium can be given daily for a few days. Alternatively, a chelated selenium product for livestock, like selenomethionine, can be given. A pinch placed directly in the mouth of the young animal daily for a couple of days can make a huge difference.

When supplemental milk is necessary, a real milk recipe is much more nutritious and healthy. There are recipes available for raising orphaned animals that use real milk, cream and a natural source of probiotics, like egg whites or yogurt. These formulas can be a safe, nutritious substitution for the commercial formulas.

One formula for a lamb or goat kid if mother's milk or a surrogate is not available

3/4 quart whole cow milk (preferably unpasteurized)
1/4 quart heavy cream (preferably unpasteurized)
1 raw whole egg beaten (farm-fresh, if available)

Modifications of this recipe exist online, but I was unable to find an original author to whom it should be attributed. The above version is a recipe I have used successfully.

CHAPTER 12

ORGANS OF DETOXIFICATION AND ELIMINATION

Understanding normal function of organs and vessels involved in the process of bringing nutrients to body cells and removing waste products allows us to better choose remedies for supporting these actions and the other organs related to this process. The following information is an overview of the eliminative systems of the body and relates to all mammals. While this information may be tedious to many people, a basic understanding of the process is useful. I have tried to simplify the processes (hopefully not to the point of incoherence!) allowing most everyone to get through this. The information following refers to mammals rather than birds, although there are many similarities. There are also some differences in avian physiology but for the purposes of this book, that will not be discussed.

When these organs of elimination are not functioning properly, toxins from metabolism, the environment and disease states accumulate in the body and clog channels of elimination. This can lead to a whole range of problems, including minerals accumulating to toxic levels or the body unable to use key minerals properly. Toxins cause irritation and inflammation, setting up the body for disease states like cancers. There are some practitioners who feel that homeopathic remedies may be less effective if the eliminative systems are not functioning well or key antioxidant minerals are not available. Understanding how these organs work to the benefit of the body enables you to recognize states of toxicity or poor functioning and gives you the option to address this using herbs or homeopathy to help stimulate elimination of toxins. Detoxification will be explained in Volume 2 but the information on organ function is given here.

Lungs

The lungs are the organs by which the body obtains oxygen and is able to remove carbon dioxide, nitrogen and metabolic water. Metabolic water is a product of the breakdown of food into component nutrients and waste products. Substances in the body that are gaseous are dissolved in water on inspiration and moved throughout and out of the body by movement of this water vapor. The blood removes unneeded dissolved gases from tissues and circulates near the alveoli of the lungs. Pressure differences between the alveoli (lower pressure) and the blood (higher pressure) allow gases to move into the lungs for transport outside the body during exhalation. Incoming air is higher in oxygen while dissolved gases in the bloodstream are higher in carbon dioxide. The exchange of gases allows blood to be recharged with oxygen, and carbon dioxide to be removed by exhalation. Contraction and relaxation of abdominal and respiratory muscles allows lungs to expand and return to normal, changing pressure and moving air in and out of the body. This cycle of respiration is very similar in all animals except horses. Horses have two phases of inspiration and two phases of expiration.

A note about breathing as it relates to pain: when an animal is relaxed and comfortable, abdominal breathing is normal and the abdomen will move with each breath. If an animal is in pain associated with the abdomen, however, the breathing may become costal or be characterized by rib movements rather than abdominal movement. If the pain is in the thorax, abdominal breathing will occur even though the animal is in pain. When assessing an animal for pain, check the breathing.

The nostrils are the beginning of the respiratory system and are usually able to dilate to allow more airflow when necessary. Inside the nostrils (nasal cavity) is a mucus surface to help filter out particles and warm and moisturize the incoming air. Blood flow through the nose also allows for cooling of the blood that goes into the brain. During extreme cold, animals and people instinctively breath through their mouths and this may help blood to the brain remain warmer than it would if the colder environmental air was in contact with the blood. Air flows from the nostrils through the pharynx and then through the larynx, where the vocal chords are located in mammals. From there, air passes through the trachea (rather than the esophagus that food travels through) and continues into the lungs.

In the lungs, the alveoli are where gas transfer occurs between air and blood. Blood comes from the heart (through pulmonary arteries) where it is called venous blood and is high in carbon dioxide and lower in oxygen. After gas transfer occurs, the blood (now bright red and high in oxygen) returns to the heart as arterial blood, passing through pulmonary veins, and is pumped throughout the body.

Carbon dioxide, in order to be transported, is most often in the form of bicarbonate in the blood. Addition of water (H_2O) to the carbon dioxide (CO_2) creates bicarbonate (HCO_3-) and a free hydrogen ion. This occurs mainly in the red blood cells with the enzyme carbonic anhydrase. In order for the bicarbonate to be removed from the red blood cells, the free hydrogen ($H+$) ion is buffered and the most common buffer is hemoglobin. In times of anemia and low hemoglobin, there is an increase in $H+$ ions and acidemia occurs.

During kidney failure, ammonia can also be released from the lungs and is detectable on the breath.

(Reece, 2009, pp. 269-291; Swenson, 1984, pp. 226-235).

Circulatory/Cardiovascular System

In a developing fetus, exchange of nutrients and wastes occurs by diffusion into and out of the uterine fluids surrounding the growing fetus. As the fetus enlarges, the distance from the innermost cells to the outside of the fetus is too far for simple diffusion and the circulatory system begins to develop to move nutrients and wastes throughout the fetus body. At its most simple, the circulatory system consists of a pump (the heart), vessels (arteries, veins, capillaries) and the fluid needed for transport of nutrients and waste (blood). The lymphatic system is a parallel system of drainage that helps return fluid to the circulatory system. Lymph is not moved by action of the heart but rather the contraction of muscles that pushes the lymph up the body.

The heart is surrounded by a sac, the pericardium, which usually contains some fluid for protection of the heart as it beats. The pericardium, however, can become inflamed and fluid-filled, causing problems. The heart itself is divided into two sides, each of which has a top and bottom chamber, atrium (top) and ventricle (bottom). The atria of each side receive the blood coming in from the veins, while the ventricles pump blood to the lungs and aorta. Between these atrium and ventricles are valves that keep blood from flowing backward into the chamber from which it was just moved.

Blood flow in the body and through the heart follows a particular path. Blood from the body comes back to the heart through the large veins: the cranial vena cava (called superior vena cava in humans) and caudal vena cava (called inferior vena cava in humans). Blood from the kidneys empties directly into the caudal vena cava. Blood in the veins is called veinous blood and it has higher carbon dioxide but lower oxygen. This carbon dioxide-rich blood is heading for the lungs by way of the right atrium, through the right ventricle, and then through a valve before entering pulmonary arteries and to the alveoli in the lungs. Pulmonary arteries are called arteries rather than veins because they are transporting blood away from the heart. Veins bring blood back to the heart, even though in most cases veinous blood is low in oxygen. Pulmonary veins carry the oxygenated blood from lungs to the heart where it enters the left atrium then into the left ventricle. The left ventricle pumps the blood through the aorta to the whole body.

The blood vessels leaving and entering the heart are large, carrying high volumes of blood to and from the body. But these large vessels cannot allow nutrients and wastes to pass from them into tissues. Instead, larger vessels (arteries) connect to smaller vessels (arterioles) and even smaller vessels (capillaries). Capillaries are very tiny vessels, as small as five to ten micrometers wide. In the walls of the capillaries are intercellular clefts (small slits) that allow transport of nutrients and wastes from the capillaries to tissue plasma and back again. These clefts are so small that larger molecules, like proteins, cannot cross. Instead, protein molecules may cross the vessel walls by vesicles in the vessel linings. Returning blood now flows from capillaries to slightly larger venules and from venules to veins for the return trip to the heart. Veins contain valves so that blood cannot flow backward, since there is no strong pumping action of the heart for returning blood flow to keep the flow moving forward.

Returning to the discussion of capillaries and their importance in transporting substances from tissue plasma brings us to the lymphatic system. Lymph vessels start within the tissue and are filled with the plasma that contains the larger protein molecules that have no way to return into the capillaries. This interstitial fluid is now called lymph. Lymph is being returned to the circulatory system via a slightly different route since it cannot re-enter the capillaries because of the larger protein molecules. Lymph vessels with their start in tissue as open-ended vessels are perfect for transport of the larger molecules and they often parallel capillaries. Like veins, lymphatic vessels have valves to prevent fluid from flowing backward.

Lymph flows into nodes at various spots in the body. Contained in these nodes are an important part of the immune system: lymphocytes that either produce antibodies or are sensitized. These lymphocytes travel in the lymph and are moved throughout the body where they attack substances foreign to the body, like bacteria and cellular waste. Lymphocytes increase in number when these substances are detected and ensnare them, leading to swelling and inflammation in the lymph node. Cancer cells moving throughout the body are also ensnared in the lymph nodes and can then increase in numbers, causing cancer of the lymph nodes where the cells were entrapped.

Lymph can only move upward toward the heart by contraction of surrounding muscles and of the lymph vessels. Because of this system of passive return, lymph and interstitial fluid can easily accumulate in extremities where lymph flow is hindered. This is one reason exercise is so important, for animals, as well as people. When a person or an animal is unable to move on their own, gentle massage designed to move lymph and fluids should be done to keep circulation and movement of nutrients and toxins into and out of cells and the body.

(Reece, 2009, pp. 228-242).

Spleen

This organ is often overlooked but is very important in filtering blood and lymph. The spleen has both red and white pulp, the red pulp is filled with blood and is filtered through macrophages. Macrophages are immune system cells that destroy old and abnormal red blood cells and recycle iron that is contained in the red blood cells. The white pulp is lymph nodes that produce lymphocytes. Blood enters either one side or the other. Because the spleen stores red blood cells, it is useful when more red blood cells are needed by the body. Contraction of the spleen releases these cells back into circulation.

(Reece, 2009, pp. 244-245).

Kidneys

The urinary system is involved not only in removing wastes (generally metabolic) but also in regulation of the body's environment and fluid levels. Kidneys eliminate urine and urea along with other waste, and remove or recycle water from the body. In order to function properly, the kidneys need salt (NaCl) and potassium and plenty of fresh, clean water.

There are three basic parts to the filtration of blood and elimination of urine. First the glomerulus, a cluster of capillaries, filters blood coming into the kidneys. Tubules then reabsorb water from the filtrate and, lastly, urine is excreted. A more detailed explanation follows.

Blood comes into the kidneys through the renal artery and is removed from the kidneys by renal veins. The renal artery comes directly from the aorta and the renal veins empty into the caudal vena cava, the largest vein in the body. Inside the kidneys are convoluted loops of tubes and associated structures, called nephrons. Different parts of these nephrons have different names. The glomerulus is a small branching of capillaries at the head of these nephrons where filtration of blood takes place. A small artery, the afferent arteriole, transports the incoming blood from the renal artery to the glomerulus. From the glomerulus, the fluid part of the blood (called plasma), minus larger protein molecules and ions unable to cross the membrane into the glomerulus, continues through the convoluted loops. The efferent arteriole moves the blood away from the glomerulus while the plasma continues through the nephrons to be further filtered. The substances removed from the blood at the glomerulus are stored in the Bowman capsule, a sac containing the glomerulus. Filtrate stored in the Bowman capsule is later moved through the proximal tubule and on to the loop of Henle. The loop of Henle is a U-shaped tubule preceded by the proximal tubule and followed by the distal tubule. Its functioning is explained in greater detail below. After much collection into progressively larger ducts, the waste from this filtration process is emptied into the bladder.

Filtration in the kidney depends on the rate at which blood enters and plasma (the liquid portion of blood) is processed through the glomerulus. The rate at which fluid moves through the loop of Henle on the way to the bladder determines how many sodium ($Na+$) and chloride ($Cl-$) ions are recycled back into the body. Flow rates are determined by release of enzymes, like renin, that affect the release of angiotensin I, a very powerful vasoconstrictor controlling the constriction of blood vessels. Angiotensin II (converted from angiotensin I) controls the release of aldosterone, an adrenal hormone. This is one way in which the adrenal glands and kidney function are related. Aldosterone helps control the reabsorption of sodium (Na+) and potassium (K+) levels in the body.

Substances that are small in size, like glucose and amino acids, can easily be eliminated from the body in urine after passing through the glomerulus. In order to move these substances back into the body, a Na+ is attached to them. The loop of Henle works on an electrical gradient between Na+ and Cl- ions to transport back and forth across membranes. Salt is vital to proper kidney function. Once glucose and amino acids are moved back into capillaries, water is left and this can either be reabsorbed or removed from the body. Potassium ions can also be reabsorbed with the Na+ ions. Ammonia (NH3) is also secreted from the kidneys, depending on the acid/base balance in the body fluids. These substances that need to be attached to a carrier before being reabsorbed by the body can only be reabsorbed at a certain rate. If that rate is exceeded, the substances will be removed in urine. In type 2 diabetes, for instance, insulin resistance in cells means blood glucose levels rise and exceed this rate of absorption. The spillover glucose is eliminated in the urine and can be used as a test for diabetes. Because the glucose is in excess, it increases osmotic pressure in the tubules and more water is retained. More water then has to be eliminated to return to balance and the animal or person must drink more to stay hydrated.

The system of transport of molecules in the kidneys is passive for much of its length and relies on diffusion. In the loop of Henle, water, salt and urea are either reabsorbed or continue down the tubes to be excreted as urine. Whether or not a substance is passed through the walls and back into the vasa recta (capillaries that are parallel to the loop of Henle) depends on the concentration of that substance in the surrounding vasa recta. Think of the loop as an "S" turned on its side so that the top becomes the left side. Fluid comes down the tube, is high in water and also contains salt (NaCl) and urea. Surrounding tissue is high in salt, however, and the permeable walls of this tube allow the water to diffuse out. The remaining fluid now starts back up the next section of the loop (the middle of the S). Here, the salt diffuses out into the surrounding tissue because that tissue now is higher in water. Molecules move to the space with lower amounts of those molecules. Towards the top of this section of tube, transport becomes active and salt continues to move out of the tube. Urea continues across the top and down the last leg of the loop where water is also moved out. This section of the loop is more open due to the actions of the anti-diuretic hormone (ADH) that is antagonistic to aldosterone (see above). Urine becomes more concentrated, consisting of the salt, urea and some water, and is now ready for storage in the bladder until elimination from the body.

The ADH is secreted by the pituitary gland and circulates in the blood to the kidneys. The HPA axis (hypothalamus/pituitary/adrenal) is a critically important gland system for proper immune function and elimination of toxins.

The permeability of the tubes in the kidneys depends on the amount of ADH, and osmoreceptor cells in the hypothalamus keep track of the amount of water in the system. In reality, these cells detect Na+ but not urea.

When there is not enough ADH, or sometimes none at all, the condition is called diabetes insipidus (not the same as diabetes mellitus mentioned above as Type 2 diabetes). In this case, the excess water cannot be removed from the urine, large amounts of very dilute urine are eliminated and thirst is increased.

As an animal ages, the kidney function can decline. When this occurs, often the input of blood to be filtered overwhelms the remaining functional tissue and fluid will build up. This edema can be particularly troublesome in the lungs but is also visible in extremities and abdomen.

There is a feedback between heart function and kidney function. The amount of the blood flow from the left atrium of the heart is dependent on, and determines, kidney function. Nerve cells in the atrium react to blood volume and control water reabsorption in the kidneys.

In addition to all of this, the kidneys are important for other reasons. In adult mammals, the kidney is the largest site of production (the only site in dogs) of a hormone, erythropoietin EPO, needed to stimulate new red blood cell production in the bone marrow. The liver is another site where this hormone is produced. Because the kidneys are the only place dogs make this hormone, when they experience kidney failure, there is often an associated anemia from lack of red blood cell production.

The parathyroid glands, located behind the thyroid gland in the neck, release parathyroid hormone that acts on the kidneys to increase reabsorption of calcium ($Ca+2$) and excretion of phosphorus. As calcium levels decrease in the kidneys, this stimulates the formation in the kidneys of calcitriol, the active form of vitamin D.

Another connection to other eliminative systems involves bilirubin, a byproduct of the liver's breakdown of hemoglobin. This is usually removed by the liver, through the gallbladder and into the intestines where it is eliminated from the body. Some of the bilirubin, however, is reabsorbed as urobilirubin and taken by the blood to the kidneys for elimination from the bladder as urine. Bilirubin is partially responsible for the yellow color of the urine when it is oxidized (exposed to oxygen) and becomes a yellowish color.

Odor of urine depends most likely on diet and kidney function. Consistency is usually water-like except in horses where a thicker urine is normal. The thick, mucus-like urine in horses is a result of mucus produced to protect the urinary system from the higher precipitated (solid) carbonates and phosphates excreted by horses.

The H+ ion mentioned above under the Cardiovascular system section is important to maintaining the pH of blood and is also important in kidney function and in the workings of the lungs. H+ is secreted in the kidneys and reacts with bicarbonate from the blood to form carbonic acid (H_2CO_3). The carbon dioxide (CO_2) and water become part of the urine.

By now it should be obvious that the intricate workings of the kidneys are important in many functions throughout the body and that their function works on a feedback system in relation to every other body system.

When the kidneys are not working properly, other organ systems cannot continue to function adequately. For instance, the lungs are forced to eliminate excess ammonia when kidney function decreases.

(Reece, 2009, pp. 313–329, 330–331, 334, 336–338, 344–345).

Skin

The skin is the largest organ of the body and serves several important functions. It protects the body from dehydration or desiccation and is the first line of defense of the body against invaders. Skin also provides a means for the animal to know what is going on outside the body using tactile sense, nerves and receptors to warmth and cold. Thickness depends on the species, age and gender and can vary from one part of the body to another.

The skin is not just one layer but made up of several layers of different types of cells and, in most animals, also has an outer covering of some type of hair. Hair provides additional protection from the sun, temperature extremes and wounds. The epidermis is the outermost layer and is made up of a thickened layer of dead keratinized cells and an inner layer of living cells produced by basal cells. The dermis contains the connective tissue of the skin, and the epidermis exists on the dermis.

The epidermis protects the body from many substances but it is also permeable to some substances, good and bad. There are two routes to permeability in the skin: transepidermal absorption, and absorption through the sebaceous glands. The sebaceous glands secrete waxy and oily substances that either protect skin or coat hairs (depending on where these glands occur). The transepidermal layer protects well against water and electrolytes (water-soluble substances) but the sebaceous glands can absorb these substances. Follicular epithelium (near the hair follicle) can also be a source of absorption and for those animals covered in a lot of hair, this can be a large source of absorption of some substances.

Substances that are soluble in fats, or lipids, more easily and quickly penetrate through these layers. Water-soluble substances can also be absorbed as water is taken in by cell membrane proteins. Substances that are soluble in lipids and water are the easiest and most quickly absorbed.

The higher the temperature (without causing damage to the animal), the more quickly the substance can be absorbed. Gases may be absorbed most quickly at higher concentrations and this may relate to the blood flow in the skin.

Breaks in the skin allow substances to penetrate more easily and inflammation or extreme hydration of the skin can also increase absorption of substances.

Why is this important to know when considering the impact of toxins and their elimination? Skin's permeability to potentially toxic substances means that contact with toxins in the external environment may lead to increased absorption and retention of toxins in the body. Some of the more important substances to consider are the lipid-soluble substances, and these include essential oils. Since the essential oils are volatile, lipid-soluble substances, they are very easily absorbed by the skin and hair follicles. In animals, the presence of hair means the route by which oils are absorbed is even greater and this may increase risk for toxicity. This will be covered more in depth in a later volume.

Lipids and water can be absorbed through the skin at rates that are similar to digestion. Some examples of lipid-soluble substances include the essential oils, salicylic acid, phenols, ketones, fat-soluble vitamins A, D, E and K, and the sex hormones progesterone, testosterone and estrogens. Salicylic acid is found in *Salix* family plants, like willow, and many other plant species (Swenson, 1984, p. 541).

Heavy metals can also be absorbed through the skin. Mercury, lead, copper, tin, antimony and bismuth are examples. Most all gases can be absorbed through the skin, with the exception of carbon monoxide. Examples of absorbable gases include oxygen, carbon dioxide, hydrogen cyanide, ammonia, nitrobenzene, dinitrotoluene (the precursor to TNT), and of course, the volatile oils.

Sunlight is famously absorbed through skin, although skin develops coping mechanisms for extreme sun exposure by producing pigments that help protect skin. Animals' skin is generally better protected from the sun by hair and pigments than human skin, although this certainly varies by species of animal. Some plants and metabolites from illness or toxins can increase the sensitivity of the skin to the effects of sun, causing sunburn in affected animals and humans. It is important to remember that although sunlight can cause damage to skin, overall it is beneficial and needed for the formation of vitamin D.

The skin, in addition to sebaceous glands, also contains sweat glands. These glands are primarily for reducing body heat, although in many animals, panting is also a large part of the cooling process. Sweat glands occur all over the body in most species and domestic animals that have sweat glands include dogs, cats, horses, pigs, sheep and cattle. Some smaller mammals lack sweat glands, like the mouse. Heat is not the only stimulation for sweat glands since the glands also perform other functions, like releasing scent and pheromones. These glands produce more than just water, however, and the secretion can be very similar to urine, although much more dilute. In this way, skin removes toxins through secretions of the sweat glands. Urea and other waste products can be eliminated through the skin when organs like the liver or kidneys become overburdened. This may cause skin conditions, as pores become plugged and toxic secretions irritate skin. Any time there are skin problems, it is always safe to assume the body has more toxins than it can eliminate through other channels.

Sebaceous glands, as noted above, are often associated with the hair follicle and secrete sebum, a waxy substance. In some species, these glands are used for extreme scent-marking and in goats, the scent glands are modified sebaceous glands located at the base of the horns. In males during rut (breeding season), the odor can be overpowering (and smelled for a quarter mile or more). Unlike sweat glands, there is no nervous control of sebaceous glands, hormones instead control production of sebum. In males, testosterone is the controlling hormone and in females it is adrenal androgens that control production of sebum. Sebum is important because it not only gives an oily shine to hair, it also acts as a fungicide, is antibacterial and plays a role in pheromones.

(Swenson, 1984, pp. 537-545).

Liver

The liver is the largest gland in the body and performs several functions. The liver filters and detoxifies blood from the body and digestive system that comes in from the portal vein from the digestive system. The detoxified blood then enters the hepatic veins and into the caudal vena cava and back to the heart. Macrophages in the liver remove foreign substances and old tissue, like broken down red blood cells.

The liver also produces bile and bilirubin that are secreted into the small intestine for digestive function and to help maintain the proper pH in the small intestine. The bile is composed primarily of bile salts that are made from cholesterol. This is the primary way cholesterol is removed from the body. Bilirubin is composed of hemoglobin that has been broken down. The intestinal bacteria convert bilirubin into urobilirubin, part of which is reabsorbed into the bile while about five percent is excreted in urine. The urobilirubin that is removed via the intestines is what causes the color of feces.

The liver and muscles also store glycogen for use in energy reserves in the body. In addition to all this, the liver stores minerals and vitamins for future use in metabolic processes in the body. A large percent of the body's vitamin A is stored in the liver. Most of the rest is stored in the adrenals, kidneys, lungs and blood (McDowell, 2000, p. 27). Because the liver stores such high levels of vitamin A, when feeding carnivores it is important not to use liver as a large percentage of their food source to decease risk of vitamin A toxicity, not to mention the increased concentration of toxins in the liver being fed (McDowell, 2000, pp. 78-79). Vitamin B12 is also stored in the liver and the discovery of the interaction between digestion and the liver in preventing pernicious anemia led to the discovery of the existence of this vitamin, based on cobalt (McDowell, 2000, pp. 523-525). Vitamin B12 is not available in plants but must be made by bacteria in the digestive system. See the section on Cobalt for more information.

The liver lobule is the basic structure of the liver and is situated around a vein. The bile canaliculi, or small grooves that are tube-like, are located near the vein and collect bile for storage in the gallbladder. The exceptions to this process are horses and a few other animals, which have no gallbladder. Quantity of bile and its constituents and concentration depends on species. Bile that is stored in a gallbladder can be more concentrated and contain other substances, like NCl, $NaHCO_3$ and water. Carnivores store very concentrated bile in the gallbladder for occasional release as food enters the small intestine. In herbivores, bile is more dilute and empties into the small intestine continuously. In horses, the hepatic bile enters the intestine directly.

The liver is also responsible for removing or recycling hormones, like the steroid hormones, cholesterol, estrogen, etc.

When the body needs vitamins and minerals that are not available from the diet, it can pull stores from the liver back into circulation again. The body cannot remove cobalt, however, from the liver for use in the digestive system. How these minerals are processed and stored depends on the vitamin or mineral. An example of mineral absorption and storage of iron shows how complex this can be.

How much iron is absorbed from the intestine during digestion depends on the body's need for iron, and this depends on how much is already stored. If enough is stored to meet the needs of red blood cell production, absorption of iron from the intestine is decreased and ingested iron is eliminated from the body. If body stores are inadequate to meet the needs of red blood cell production, however, more iron is absorbed. When more iron is taken in than needed and cannot be eliminated, iron will be stored in body cells, especially the liver, and may build up to toxic levels. The storage form of iron in the body is ferritin while a more soluble form, hemosiderin can accumulate when there is an excess (Reece, 2009, p. 63). See section on Iron in Minerals.

Storage of toxins and minerals can be in the liver itself or in adipose (fat) tissue of the body. Minerals not needed by the body immediately, or heavy metals, are often stored in the liver. This is why testing liver tissue for trace minerals and metals can give a good idea of how the body has been using those substances over the last three to six months. When the liver is overburdened with toxins, it can no longer effectively filter the blood. Toxins build up and other eliminative systems try to compensate for this build up. Hormone levels become unbalanced and reproduction, adrenal function and other body systems start to malfunction (Reece, 2009, p. 379; Swenson, 1984, pp. 298-299).

EPILOGUE

This volume touches on several topics related to raising healthy animals on healthy land. The information that could be added is immense, but more important than another book filled with information is fostering an understanding of how to adapt feeding to the needs of different species, recognizing ill health and its causes and adapting feeding and farming methods to specific species and local ecosystems.

Some books will advise that soil and forage tests be done with the intent to add amendments to soils in macro amounts. My thoughts on this have slowly changed over time and with experience. I'm not interested in making my farm conform to the standardized soils in laboratory analysis. I want to adapt my way of living and raising animals to conform to the ecosystem in which I live.

Part of this involves making conscious decisions about species and breeds of livestock that fit my goals and the ecosystem in which they will be raised. Another part involves choosing what areas of the farm will be managed as pasture or hay and what will be left as native elements to promote wildlife and intrinsic value (erosion control and filtering of water, habitat, beauty, variety of plant and animal communities).

Using rotational grazing can mimic the movements of herbivores across vast expanses of open range. This is better for the animals and better for the environment. It also allows animals to graze higher up, thus avoiding most parasite contamination.

Pasture diversity is important to health of the animals and the soils. Some plant roots reach deep, some plants provide higher nutrition and different mineral content during differing times of the growing season. Encouraging as much diversity of plant species and communities as possible increases the nutrition animals can get from their grazing and maximizes the use of the pasture. Healthy herbivores feed healthy carnivores as well.

I encourage everyone interested in having animals as part of their lives to consider which animals are better adapted to each situation, how those animals will fit into any area where they reside and then decide how best to meet the nutritional, emotional and physical needs of those animals. Work with the land and the animals and you will reap the reward of knowing a lifelong relationship with a complex and beautiful world, its inhabitants and the Creator who graciously allows us to live in it.

RESOURCES IN THE U.S.

Herbs

Back in Balance Minerals®
Back in Balance Blends
backinbalanceminerals.com
877-487-6040

Mountain Rose Herbs-dried bulk herbs, extracts, essential oils, containers and other products
mountainroseherbs.com
800-879-3337

Woodland Essence-herbal extracts, flower essences
woodlandessence.com
315-845-1515

Frontier Co-op-bulk herbs
frontiercoop.com
800-669-3275

Essential Oils

Original Swiss Aromatics
originalswissaromatics.com
415-459-3998

Mountain Rose Herbs-dried bulk herbs, extracts, essential oils, containers and other products
mountainroseherbs.com
800-879-3337

Mineral Supplements

Back in Balance Minerals®
backinbalanceminerals.com
877-487-6040

North Central Feed Products, LLC
877-487-6040

HyViewFeeds
hyviewfeeds.com
507-493-5564

Flower Essences

Flower Essence Society
flowersociety.org
800-736-9222

The Original Bach Flower Remedies
bachflower.com
800-214-2850

INDEX

J

jaundice 85, 167, 168, 188, 190, 211
jejunum 21, 45, 105, 108, 117
Juglans nigra 219
Juniperus virginiana 213, 219

K

kale 166, 220
kelp 39, 55, 62, 87, 140, 160, 163, 218, 219, 221, 222, 230, 281
keratin 165
ketones 108, 255
kid 17, 18, 58, 65, 84, 94, 98, 175, 217, 241, 242
kidneys 46, 48, 54, 69, 74, 77, 86, 91, 93, 94, 97, 105, 108, 112, 113, 117, 122, 126, 130, 137, 148, 149, 150, 151, 153, 154, 158, 161, 163, 164, 173, 178, 179, 188, 190, 192, 198, 201, 216, 230, 234, 245, 246, 248, 249, 250, 251, 252, 253, 256, 257
Krebs cycle 94, 101, 108, 208

L

lacrimation 70, 98, 110, 167
lactation 58, 74, 87, 94, 95, 105, 118, 123, 141, 142, 145, 149, 153, 161, 201, 212
lactoferrin 208
LaMancha 29, 30
lamb 8, 18, 36, 58, 59, 65, 74, 84, 94, 96, 98, 103, 124, 127, 141, 147, 174, 175, 176, 184, 185, 187, 198, 217, 225, 227, 238, 241, 242, 281
lambs quarter 147
Laminariales 39
laminitis 21, 28, 148, 149

large Intestine 10, 11, 21, 22, 33, 34, 45, 105
larynx 245
legumes 38, 71, 124, 147, 161, 181, 197, 217, 222, 233, 236
lemons 129
leucocytes 224
L-gulonolactone 132
limes 129, 223
lingonberry 213
lipid 173, 254, 255
Liquidambar stryaciflua 219
liver 6, 10, 14, 20, 21, 22, 45, 55, 56, 63, 65, 66, 69, 71, 72, 73, 75, 76, 77, 79, 80, 82, 83, 85, 89, 90, 92, 93, 94, 97, 100, 105, 106, 108, 110, 111, 112, 113, 116, 117, 122, 124, 125, 126, 173, 178, 179, 182, 185, 187, 188, 189, 190, 191, 192, 193, 194, 197, 199, 201, 203, 205, 207, 208, 209, 210, 211, 216, 224, 225, 228, 230, 241, 252, 256, 257, 258
liver alcohol dehydrogenase 224
Lotus corniculatus 38
lungs 69, 70, 77, 108, 168, 211, 224, 244, 245, 246, 251, 252, 253, 257
lymph 52, 69, 77, 246, 247, 248
lymphatic 246, 247
lymphocytes 75, 106, 247, 248
lymphocytic thyroiditis 219
lysine 101, 120, 130, 132

M

macrophages 248, 256
magnesium 21, 54, 57, 58, 61, 65, 101, 140, 141, 142, 143, 144, 145, 147, 148, 149, 150, 151, 152, 155, 156, 157, 160, 162,

omnivore 2, 9, 11, 12, 18, 20, 21, 46

osmoreceptor 251

osteomalacia 75, 76, 143

osteoporosis 74, 143, 144, 157, 185, 221, 234

ovaries 75, 155, 216

oxalates 28, 143

oxalic acid 28

oxygen 24, 70, 73, 133, 208, 222, 244, 245, 246, 252, 255

P

PABA 116, 117

paddling 149

p-aminobenzoic acid (PABA) 116, 117

pancreas 10, 14, 20, 21, 22, 27, 45, 69, 75, 77, 118, 210, 227

pancreatic carboxypeptidase 224

pantothenic acid 94, 108, 109, 110, 111, 112, 123

papillae 19, 49

parakeratosis 225, 229, 231

parasites 3, 4, 26, 29, 50, 58, 59, 60, 64, 77, 84, 86, 124, 139, 164, 166, 167, 171, 180, 181, 182, 185, 193, 197, 201, 208, 210, 212, 213, 237, 259

parathyroid glands 74, 75, 77, 146, 252

parathyroid hormone 73, 74, 142, 154, 252

parotid 16, 42, 45, 49

parsley 65, 213

pasteurization 240

pasture vii, 5, 6, 7, 8, 26, 28, 32, 33, 36, 37, 38, 55, 56, 57, 62, 63, 68, 77, 78, 79, 86, 90, 99, 100, 101, 102, 112, 118, 120, 135, 136, 138, 139, 140, 145, 147, 151, 152, 156, 161, 162,

163, 170, 171, 175, 176, 177, 180, 196, 200, 203, 206, 209, 212, 213, 218, 220, 231, 235, 236, 259

pepsin 32, 42, 121, 122

perchlorates 217

pericardium 246

pernicious anemia 63, 121, 124, 184, 193, 210, 257

peroxide 131, 173

persimmon 231

pesticide 104, 206, 214, 222

Petroselinum crispum 65, 213

Phalaris 125

pharynx 10, 245

phenylalanine 122

pheromones 255, 256

phosphates 56, 82, 94, 97, 101, 104, 105, 127, 144, 153, 155, 204, 221, 225, 252

phospholipids 153

phosphoproteins 153

phosphorus 56, 57, 61, 74, 75, 76, 140, 141, 142, 143, 144, 145, 146, 147, 148, 151, 152, 153, 154, 155, 156, 157, 161, 170, 178, 203, 204, 205, 206, 210, 222, 225, 234, 235, 252, 280

phosphorylation 161

photosensitivity 124

phytase 145, 146, 153, 156, 157

phytates 143, 145, 153, 156, 157, 203, 209, 210, 222, 223, 228, 229

pica 228

pigeons 7

pigs 12, 20, 22, 23, 64, 82, 85, 91, 93, 95, 99, 103, 105, 109, 110, 111, 112, 114, 116, 119, 125, 128, 130, 132, 134, 135, 145, 146, 151, 153, 155, 156, 157, 160, 164, 166, 168, 178, 190,

REFERENCES

Abrams, J. (2000). *Linton's Animal Nutrition and Veterinary Dietetics.* India: Greenworld Publishers

Alfs, M. (2003). *300 Herbs: Their Indications & Contraindications.* MN: Old Theology Book House.

Ammerman, C., Baker, D. and Lewis, A. (1995). *Bioavailability of Nutrients for Animals.* CA: Academic Press.

Arthington, J. (2003). *Mineral Antagonisms May Influence Copper Deficiencies.* Feedstuffs: Nutrition & Health/Beef June, 16, 2003 pp 11-12.

Bebbington, A. (2013). *Toxicity of Equisetum to Horses.* http://www. omafra.gov.on.ca/ english/livestock/horses/facts/07-037.htm Retrieved 7/15/2014

Bella, S., Grilli, E., Cataldo, M., & Petrosillo, N. (2010). *Selenium Deficiency and HIV Infection.* Infectious Disease Report. 2010 Aug 4; 2(2)

Carlson, D. & Giffin, J. (1980). *Dog owner's home veterinary handbook.* NY: Howell Book House.

Colburn, M. (2007). *Equine Health as Seen Through the Eye* (2nd ed). CA: Through the Eye International

Coleby, P. (2006). *Natural Sheep Care.* TX: AcresUSA

Coleby, P. (2010). *Natural Horse Care.* TX: AcresUSA

Curtis, L. (2011). *Feline Nutrition.* USA: Curtis.

Dworkin, B. (1994). *Selenium deficiency in HIV infection and the acquired immunodeficiency syndrome (AIDS).* Chemico-Biological Interactions Jun; 91(2-3):181-6.

Elahi, S. et al. (2013). *Immunosuppressive CD71+ erythroid cells compromise neonatal host defence against infection.* Nature 504, *pp. 158–162*

Erickson, A. (2015). *Cobalt Deficiency in Sheep and Cattle.* Aus: Department of Agriculture and Food. Retrieved 12/3/2015 https://www. agric.wa.gov.au/livestock- biosecurity/cobalt-deficiency-sheep-and-cattle

Feedstuffs: Nutrition & Health/Beef June, 16, 2003 pp. 11-12.

Fischer-Rizzi, S. (1996). *Medicine of the Earth.* OR: Rudra Press

Fluharty, F. *Updating Phosphorus Supplementation in Ruminants to Meet the Animal's Requirement, Reduce Excess Cost, and Reduce Environmental Concerns.* Dept. of Animal Sciences: Ohio State University

Fowler, M. (2010). *Medicine and Surgery of Camelids.* IA: Wiley-Blackwell.

Fox, J. (1988). *Biology and diseases of the ferret.* PA: Lea & Febiger.

Fyles, F. 1919. *Principal Poisonous Plants of Canada.* Dominion of Canada.

Getty, J. (2013). *Equine digestion, it's decidedly different.* USA: Getty.

Grieve, M. (1971). *A Modern Herbal, Vol. 1.* NY: Dover.

Hourdebaigt, J. (2007). *Equine Massage,* 2nd ed. NJ: Wiley Publishing.

Jain, N. (1993). *Essentials of Veterinary Hemotology.* PA: Lea & Febiger.

Jordan, P. (2009). *Mark of the Beast, Hidden in Plain Sight.* KY: Jordan

Kainer, R. and McCracken, T. (2003). *Dog anatomy, a coloring atlas.* WY: Teton NewMedia.

Kainer, R. and McCracken, T. (2013). *Feline anatomy, a coloring atlas.* WY: Teton NewMedia.

Kilfoyle, S. & Samson, L. (1996). *Completely Angora.* Canada: Samson Angoras.

Levy, J. (1991). *The Complete Herbal Handbook for Farm and Stable.* London: Faber.

Lewington, J. (2002). *Ferret husbandry, medicine & surgery.* MA: Butterworth- Heinemann.

Merck 2011. http://www.merckvetmanual.com/mvm/index.jsp?cfile=htm/bc/213302.htm Retrieved 11/15/2011.

Macdonald, D. (1993). *The encyclopedia of mammals.* NY: Facts on File, Inc.

McDowell, L. (2000). *Vitamins in Animal and Human Nutrition.* IA: Iowa State University Press.

Miller, M., Christensen, G., Evans, H. (1965). *Anatomy of the dog.* PA: W. B. Saunders Co.

National Research Council. (1985). *Nutrient Requirements of Sheep.* Wash. DC: National Academies Press.

National Research Council. (1989). *Nutrient Requirements of Horses, 5th Ed.* Wash. DC: National Academies Press.

National Research Council. (1998). *Nutrient Requirements of Swine, 10th Ed.* Wash. DC: National Academies Press.

National Research Council. (2001). *Nutrient Requirements of Dairy Cattle, 7th Ed.* Wash. DC: National Academies Press

National Research Council. (2005). Mineral Tolerances of Animals, *2nd ed.* Wash. D.C.: National Academies Press

National Research Council. (2007). *Nutrient requirements of small ruminants.* Wash. D.C.: National Academies Press

Pet Food Industry. (2013). Statistics. Retrieved December 5, 2013 from http:// www.petfoodinstitute.org/?page=Statistics

Pitzen, D. (1993). *The Trouble with Iron.* Feed Mgmt 44(6) pp. 9-10.

Pottenger, F. (2012). *Pottenger's cats, a study in nutrition.* CA: Price-Pottenger Nutrition Foundation.

Pugh, D. (2002). *Sheep and Goat Medicine.* PA: Saunders

Radostits, O. et al. (2000). *Veterinary Medicine: A Textbook of the Diseases of Cattle, Sheep, Pigs, Goats and Horses.* W.B. Saunders.

Reece, W. (2009). *Functional Anatomy and Physiology of Domestic Animals.* IA: Blackwell Publishing.

Self, H. (2001). *A Modern Horse Herbal.* Great Britain: Kenilworth Press

Schreiber, M. (1996). *The "How To" Manual of Sports Massage for the Equine Athlete.* Candlewyck Press

Smith, M. (2009). *Goat medicine.* IA: Blackwell Publishing.

Stewart, J. (1950). *Induced Cobalt Deficiency in Lambs.* England: Moredun Institute.

Swenson, M. (1984). *Duke's Physiology of Domestic Animals* (10th ed.). NY: Cornell Univ.

Thorvin. (2013). *Icelandic Geothermal Kelp-Typical Analysis.* Retrieved 11/16/2015 http://thorvin.com/downloads/Thorvin-Pets-Asco-Kelp-Analysis.pdf

Tisserand, J., Hadjipanayiotou, M. & Gihad, E. *Digestion in goats.* Retrieved 1/8/2014 http://www.ladocumentationcaprine.net/plan/alimentation/art/artali5.pdf

Underwood, E. (1981). *The Mineral Nutrition of Livestock.* England: Commonwealth Agricultural Bureau.

Van Dyke, N. (1998). *"Interpreting a Forage Analysis"* Alabama Extension. http:// www.aces.edu/pubs/docs/A/ANR-0890/ Retrieved 7/11/2013.

Walters, C. (2013). *Minerals for the Genetic Code.* TX: AcresUSA

Wynn, S & Fougere, B. (2007). *Veterinary Herbal Medicine.* MO: Elsevier.

Zucker, M. (1999). *The veterinarian's guide to natural remedies for dogs.* NY: Three Rivers Press

ABOUT THE AUTHOR

Alethea Kenney has been farming since 1996 and has lived and farmed in northern Minnesota since 2001. Her love of all animals, particularly her dogs, has led her to devote her life to improving the health of animals everywhere. She spins, knits, weaves and does nalbinding with the fibers from her animals and uses both natural colored fiber and natural dyes to add interest to her projects.

She has designed her own bioavailable livestock mineral line, Back in Balance Minerals®, and herbal blends for health in livestock, Back in Balance Blends. Both lines of products are available through North Central Feed Products, LLC where Alethea is the small ruminant nutrition consultant. More information and ordering of these products can be found at backinbalanceminerals.com.

Alethea, a traditional naturopath, Western herbalist, aromatherapist and homeopath, writes and teaches about restoring and maintaining health in animals through her natural health consulting company, Boreal Balance, LLC. She also writes about herbs and minerals as they relate to health in humans and animals and natural dyes, fiber art and sustainable farming methods.

She is board certified by the American Council of Animal Naturopathy and has degrees in Wildlife Science (B.S. Purdue), Naturopathy with concentration in herbology (N.D. CCNH), Aromatherapy Masters series (Internationally certified through Pacific Institute of aromatherapy), certified in Equine Iridology (Through the Eye International), certified in Western Herbalism (Midwest School of Herbal Studies), Diploma in Veterinary Homeopathy (British Institute of Homeopathy).

Her articles on minerals have been published in Sheep! magazine and her article on natural dyes was published in the Journal of the American Herbalists Guild. She has also had articles published in The Shepherd magazine and Icelandic Sheep Breeders of North America newsletter.

More information can be found on her websites: borealbalance.com, backinbalanceminerals.com and reedbird.com.